T0147910

The Saga
of the
Tin Goose

The Story of the Ford Tri-Motor
3rd Edition 2012

DAVID A. WEISS

Order this book online at www.trafford.com
or email orders@trafford.com

Most Trafford titles are also available at major online book retailers.

© Copyright 2013 David A. Weiss.
All rights reserved. No part of this publication may be reproduced, stored in a retrieval
system, or transmitted, in any form or by any means, electronic, mechanical, photocopying,
recording, or otherwise, without the written prior permission of the author.

Cover Photograph U.S. Forest Service courtesy of Penn R. Stohr, Jr.,
Evergreen International Airlines.

Printed in the United States of America.

ISBN: 978-1-4669-6902-5 (sc)
ISBN: 978-1-4669-6901-8 (e)

Trafford rev. 12/21/2012

 www.trafford.com

North America & international
toll-free: 1 888 232 4444 (USA & Canada)
phone: 250 383 6864 ♦ fax: 812 355 4082

To my young eagles Kirk and Edward

Contents

Introduction

The Ford Tri-Motor—or the "Tin Goose," as it is affectionately called—is the airplane that ushered in the age of American commercial aviation.

Technologically, this plane, with its familiar corrugated-metal skin, noisy engines, and barndoor-like rudder, encompassed many aeronautical innovations: all-metal construction, the enclosed cockpit, the cantilever wing—these are but a few.

Yet, important as these innovations were, they represent only a fraction of this plane's total contribution to American aviation. More than any other airplane, the Ford Tri-Motor transformed American civil aviation from small fly-by-night operations to the giant airline industry of today. Most major United States airlines began with Ford Tri-Motors, and in these planes our nation's first transcontinental air service was inaugurated. As a direct result, the United States saw its first successful passenger airlines, its first scheduled flights, its first modern airport, and even such now-accepted airline amenities as stewardesses, in-flight meals, lavatories and pilots that wore uniforms.

Despite all these contributions, only 199 Ford Tri-Motors were ever manufactured, and the plane's reign as the nation's No.1 airliner was brief, lasting only from the late 1920s to the early 1930s. Yet, when they did fly the skies on major air-passenger routes, no other planes could touch them in dependability, speed, and size. With their all-metal construction (at a time all other American planes were constructed of wood), their capacity of some fifteen passengers (when most passenger land planes could carry no more than three or four), their then-large size (fifty feet from nose to tail), the Ford Tri-Motors were considered the last word in passenger air travel in the United States until the advent of the speedier Boeings and Douglases, which one by one supplanted the Fords on the major air routes.

The onetime "Queen of the Skies" was demoted operationally. But in the hearts and minds of the pilots who flew the planes, and the passengers who flew in them, affection and warmth replaced admiration and awe. As the Ford Model T automobile had been personified as the "Tin Lizzie," so now the Ford Tri-Motor became the "Tin Goose." The plane had other names—many like "Flying Barndoor" tacked on by journalists trying to get fanciful—but it was the Tin Goose name that stuck. It was not only the association with the Tin Lizzie name that was responsible. It was also the way the plane waddled down the runway on takeoff, strikingly similar to a goose waddling with her goslings. It was also the assortment of noises its three noisy engines made that sounded like an aeronautical honk.

As the Tin Gooses were taken off major air lanes, they began to be flown on secondary routes, and then they began flying south of the border. They flew to remote jungle plantations and mountain mines in Central and South America, landing in almost impossible terrains where larger and speedier aircraft could not operate.

On and on they flew—in Mexico, in Australia, and in China. And in the early days of World War II, a Tin Goose suddenly appeared almost out of nowhere in the Philippines to help evacuate United States troops from Bataan.

After World War II they started new careers, dusting crops and fighting fires in the west and mid-west, participating in air shows, and over the years touring country fairgrounds and municipal airports, giving millions of Americans their first ride in an airplane, and then for the few that have survived, getting well-deserved rests in one of the nation's aviation museums where they are often the prime attraction.

It's a great story.

Chapter One
In The Beginning

The Ford Tri-Motor plane was born in the post-World War I era. After the doughboys came marching home from France in 1919, a bright future was predicted for American aviation. But from the beginning, commercial aviation could not get off the ground. Until almost the mid-1920s, the industry was in a sorry state.

There were no organized airlines worthy of the name, and what companies did exist were small, precariously financed, and irregular in their operations. By and large, commercial aviation was limited to taxi flights from local airports and sightseeing tours from the bumpy airfields of county fairs.

Typical of these early airlines was the Robertson Aircraft Corporation of Saint Louis, Missouri. Organized in 1921, this company's entire capitalization was only $15,000. Its equipment consisted of one Curtiss biplane powered with a Curtiss OX-5 engine; two complete sets of engine parts; and miscellaneous spare parts. As for its operations, Robertson had a lease on an airfield site, and several contracts for flying in conjunction with advertising.

One reason for the sad condition of American aviation was the paucity of facilities. There were only a few good landing fields in the entire United States, and these were wretchedly equipped. Any type of weather information service was practically nonexistent. There were no beacons to serve as guides for routes, no radio stations, no radio beams.

Plane flights were accordingly very limited. None could be made safely in bad weather; nor could planes fly at night. As a result, not much time was saved by flying, and the public did not consider aviation practical for transporting either passengers or freight. Indeed, by 1923—exactly

Preceding Page: Jack Knight, whose heroic flight on February 22, 1921, gave the Air Mail Service a new lease on life, UNITED PRESS INTERNATIONAL PHOTO

two decades after man's first successful heavier-than-air flight—no aircraft had yet been built in the United States that could carry for any considerable distance a respectable load of passengers or cargo. The few commercial planes available were single-engine biplanes, constructed of wood and fabric, with top speeds of not more than 100 mph. And most of these were of World War I vintage, like the De Havilland DH-4, a two-seater observation plane originally flown over the Western Front by United States Army pilots.

Such planes were little more complicated than the first Wright biplanes. They were usually equipped with fewer than a dozen instruments and meters, such as compass, oil gauge, tachometer, and altimeter. For navigating, there were no aerial maps, and the only reliable navigation guide for pilots was the "Iron Compass"—the railroad tracks they could spot from above connecting one city with another. As for controlling these early airplanes, this was sometimes impossible. On one occasion, a pilot, while flying, was caught in a storm. With his plane bucking and tossing, he could not keep it on an even keel. At last, in desperation, the pilot took his hands and feet off the controls. "Here, God, you take over. I've done all I can!" he shouted. And, as he recounted later, "You know what? God cracked it up."

About all that kept aviation alive in these post-World War I days were the ex-pilot heroes who had flown in the war. Some stayed in the service and performed feats of aerial accomplishment that made headlines all over the world. In May, 1919, three United States Navy NC-4 Curtiss seaplanes took off from Long Island on what was hoped would be the first successful crossing of the Atlantic. Two of the planes were forced down at sea, but one NC-4 reached its destination in the Azores. That same year two Englishmen, John Alcock and A. W. Brown, flew through wind and fog in a Vickers "Vimy" bomber from Newfoundland to Ireland in sixteen hours. This was the first nonstop flight over the Atlantic.

Most former war pilots, however, ended up in less glamorous situations. Although flying was in their blood, a rapidly demobilized Air Corps left few planes to fly in, so many of the men settled down to peacetime aviation. Some purchased war-surplus Curtiss Jenny trainers for three hundred dollars and set up flying schools. Others, either as lone wolves or as performers in aerial circuses, became barnstormers, touring the country as gypsy pilots. From town to town they would fly, making their headquarters on a recently cleared farmer's field which, low stumps and all, served as a primitive landing field. Then, on the scheduled afternoon, they would put on their show. As crowds assembled, they would start stunting, putting their flimsy aircraft through incredible aerial maneuvers. Without parachutes they would walk the wings of their planes. Flying upside down, they would pick up handkerchiefs from the tops of waving weeds. They would jump from one plane to another in midair. They would engage in mock dogfights. And then, after the exhibitions were over, they would take up passengers at ten dollars a head, or as inexpensively as one cent per pound of the customer's weight.

The Curtiss "Jenny." Originally used to train army pilots in World War I, these planes were purchased after the war by demobilized pilots for barnstorming, PICTURE COLLECTION, NEW YORK PUBLIC LIBRARY

More often than not, these aerial vagabonds and their planes were one jump ahead of the sheriff. But they did not care; they were flying. And, as in the war, they flew as part of the plane. They "read" the sky by instinct and "smelled" the weather. Exposed to the elements in their open cockpits, they often felt rain fall on their faces, and sometimes even ice crystals formed there. But what did a little discomfort mean? They were flying.

While commercial aviation in the United States had been arrested at the barnstorming stage, commercial aviation in Europe was making significant strides. Airline companies there were backed by government subsidies. By 1921, more than ten airlines linked various European capitals. In England alone, there were five airlines carrying passengers, and in France, six.

The low state of American aviation was particularly ironic, since the first heavier-than-air powered flights had taken place in the United States. The achievement of the Wright brothers in making this possible stands as one of the great engineering advances of all time.

From earliest days, men had dreamed about flying. But in trying to fly themselves, the ancients soon came to realize that the way of the birds would not work for man. The muscles of human beings were simply too weak to achieve flight. Eventually it was realized that human flight could be accomplished by only two methods: one, by a process known as aerostation, via a lighter-than-air machine that would float in the air; two, by a process known as aerodynamics, via a heavier-than-air craft that would take advantage of reactions developed by the rapid movement of bodies through the air.

The lighter-than-air method was perfected first. As far back as the sixteenth century, Roger Bacon, a Franciscan friar, reasoned correctly that flight was possible in a sphere of fine metal filled with a lighter-than-air substance. Bacon unfortunately did not know of such a substance, but his idea took hold, and was advanced further by the seventeenth-century Jesuit de Lana, who even drew a design for a vacuum balloon consisting of four copper spheres, filled with "lighter-than-air" gases.

Orville Wright piloting the Wright brothers' third plane in 1909—the first fully practical airplane in history, UNITED PRESS INTERNATIONAL PHOTO

Then came the first lighter-than-air flight. At Annonay, France, in 1783, Joseph and Etienne Montgolfier, paper-makers, sent up a paper balloon filled with hot air. It rose six thousand feet and landed more than a mile away. Soon after came another ascension, this one before Louis XVI and the royal family, with a lamb, rooster, and duck as passengers. And before the year was out, the king's historian, de Rozier, went up in a Montgolfier balloon, becoming the first human being to ascend (and, when he crashed on a later ascension, the first aerial casualty).

Soon after, hydrogen—a gas fourteen times lighter than air—began to be used in balloons, and lighter-than-air craft developed more rapidly.

David A. Weiss

In the mid-nineteenth century, Professor Thaddeus Lowe made an eight-hundred-mile balloon voyage from Cincinnati, Ohio, to South Carolina, and before the end of the century, Count Ferdinand von Zeppelin built his first semi rigid dirigible, a ship four hundred twenty feet long. By 1914, when World War I broke out, some twenty-six Zeppelins had been constructed, and more than eight hundred flights had been made over Europe.

In heavier-than-air craft, a prime innovator was Leonardo da Vinci, painter, engineer, anatomist, scientist, and sculptor. Da Vinci had a great love for birds and he not only devoted hours to studying flight, but also roamed the streets, purchasing birds in cages just so he could set them free. As a result of his studies, da Vinci drew sketches of such aerial devices as propellers, helicopters, and parachutes (which he called "fall breakers"). His helicopter, which he designed in detail, would work successfully, he said, if only he had an engine that could turn at sufficient speed the helix—the screw-shaped spiral serving as the vehicle's propeller. (Present-day aeronautical engineers who have studied his design agree.)

The first actual step in heavier-than-air flight took place several centuries later. In the 1770s, Sir George Cayley in England constructed a number of successful wood-and-cloth gliders capable of carrying human beings, and then a working model of a helicopter-type "aerial carriage" complete with propellers and rotating wings. One of Cayley's first gliders carried his coachman safely to the ground after being launched from the crest of a hill. Another of his gliders was the first biplane glider, the second wing being added to give additional lifting power. Cayley, too, was one of the first to realize the importance of streamlining in aeronautical design. He also introduced the concept of a tail for longitudinal stability and control. So attuned to the potential of heavier-than-air flight was Cayley that he predicted, "the noble art of flying would soon be brought home to Man's convenience" and "we shall be able to transport ourselves and our families and our goods and chattels. more securely than by water, and with a velocity of from 20 to 100 miles per hour."

Next came William Henson who, in the mid-nineteenth century, thought of hooking up a small steam engine to a glider. His resulting "Ariel Steam Carriage" resembled an early monoplane, but it had no lateral stability and thus was unable to sustain itself in the air.

Gliders continued to be improved, and before the beginning of the twentieth century John J. Montgomery, a professor at Santa Clara College in California, secretly assembled with his brother a gull-like glider of sticks and bits of cloth. Launched from a hill near San Diego, the glider crashed, but another one, carrying a man aloft, flew successfully shortly afterward.

Also successful with gliders at this time was the German, Otto Lilienthal. At the age of thirteen he had begun his experiments, working, like Montgomery, with his brother. At first Lilienthal constructed frames to which he attached linen remnants, flight being attempted by flapping the frames. When this failed, he tried a machine made of glued-on wings, and again he met failure. Next, Lilienthal studied the flights of birds, particularly storks, which, he noticed, glided with motionless wings. Then he published his classic *Der Vogelflug als Gmndlage der Fliegekunst* (The Flight of Birds as the Foundation of the Art of Flying). Soon after, Lilienthal began building workable gliders, and in 1891 he successfully flew in a glider made of peeled willow saplings and cotton cloth, over which a wax coating had been applied to achieve air tightness. With this glider, Lilienthal became adept at utilizing air currents and banking. Other gliders followed, biplane gliders and gliders with flapping wing tips, and Lilienthal eventually added motive power in the form of an engine driven by compressed carbonic gas. In 1896, in one of these gliders, Lilienthal took off from a hill near Stollen. He attained a height of fifty feet, but when the wind suddenly abated, the glider nosed downward and crashed. Lilienthal's spine was broken, and he died the next day. His last words were "sacrifices must be made."

Meanwhile, in Australia, Lawrence Hargrave, an assistant at the Sydney Observatory, also turned to the problem of human flight. Studying soaring birds, he concluded that heavier-than-air flying was possible only with an engine. He had already built monoplane glider models as well as box kites. Now he developed a rotary engine powered by compressed air and designed so its cylinders revolved around a stationary crankshaft. Although this was the first practical engine designed for use in flight, Hargrave made no attempt to patent its principles. Later, when the Wrights asked to use it, he offered them permission free, saying, "My inventions and discoveries are for humanity."

Another pioneer was Octave Chanute, a French-born American who became one of the United States' most prolific engineers. Interested for many years in the problem of flight, he did not actually become active in the field until the age of sixty, when he read Lilienthal's work. This led Chanute to construct gliders, and in the 1890s he made several hundred glider flights over the sand dunes of Lake Michigan. Chanute also published a significant book titled *Progress in Flying Machines.* But his most important contribution was encouraging the Wright brothers and giving them practical suggestions about biplane construction.

Then came Professor Samuel P. Langley of the Smithsonian Institution. Intrigued by the possibilities of flight in heavier-than-air machines, Langley constructed steam-powered models called "aerodromes," which had curved lifting surfaces and flew successfully. As a result of these flights, Langley received a $50,000 grant from the War Department to construct a workable flying machine. But, as he himself realized, powered flight was impossible unless a lighter engine could be developed. This problem was solved by his assistant, Charles M. Manley, who designed a small water-cooled radial engine powered by gasoline. A small version of this engine was successfully used in a quarter-sized model of a new aerodrome. But in October, 1903, when Langley tried launching a full-sized aerodrome via catapult from a houseboat anchored in the Potomac River, he failed. In December of the same year, just two weeks before the Wrights made their first successful flights, Langley tried again and failed, after which he abandoned all future experimentation.

Langley and all the others represented only the prelude to the great feat of Orville and Wilbur Wright. From early childhood, they had been interested in flying, starting from the time their father, a United Brethren bishop, gave them a toy helicopter. As they grew to manhood they dreamed of flying and began experimenting with kites and gliders. They made their living by operating a bicycle shop in their home town of Dayton, Ohio, and here they first read Octave Chanute's *Progress in Flying Machines,* which spurred them on to try to build man-carrying gliders. By 1899 they had reached the conclusion that the key to sustained flight was the speed and volume of the wind around the glider's wings, and they decided that further testing should be carried out in a locality where winds were strong and prevailing. Studying United States Weather Bureau meteorological charts, they selected the area around Kitty Hawk, North Carolina, as the site for their glider tests.

The "Aerodrome" of Professor Samuel P. Langley ready to be launched via catapult from a houseboat anchored in the Potomac River in 1903.
PICTURE COLLECTION, NEW YORK PUBLIC LIBRARY

They journeyed there in 1900, and by watching buzzards fly, discovered an important new principle. It was that control of flight could be improved by wing "warping," i.e., twisting the wings in a spiral fashion to give more lift to one side and less to the other. This, they thought, would enable a glider to right itself after being rocked to one side by a gust of wind. The principle had worked with buzzards; they felt sure it would work with gliders, too.

The brothers' conclusions were right, and, as their experiments proved, when they combined a warping control with movements of the rudder, they could steer their gliders from the right or left and achieve perfect directional control.

Using this principle, the Wrights made hundreds of glider flights from the Kill Devil sand dunes near Kitty Hawk, and then in 1903, after further extensive research on propellers, they designed and built a crude 12 hp engine. By December, 1903, they had installed this engine on a glider with linen-covered wings, and they attempted to fly it. The test failed when the machine jumped the greased rail from which they intended to catapult the plane, and the plane's wing was broken.

On December 17, 1903, before a small audience of neighborhood fishermen, they tried again in the face of a bitter cold biting wind. The engine was started, and, sputtering and coughing, it vibrated noisily the frame of the frail craft. Seated in the glider, Orville hoisted the signal for his brother to remove the blocks from the front of the skids. Then, as Orville slipped off the release wire, the plane ran down the skids, rose eight to ten feet in the air, wavered momentarily, then steadied and rose higher, attaining a speed of 30 mph before plowing down into a nearby high sand dune.

The flight had lasted only twelve seconds, and the distance covered only 120 feet, but as Orville himself wrote later, "It was nevertheless the first in the history of the world in which a machine carrying a man had raised itself by its own power into the air in full flight, had sailed forward without the reduction of speed, and finally landed at a point as high as that from which it had started."

Other flights followed that day, Wilbur on the fourth flight flying 852 feet and staying up in the air fifty-nine seconds. However, the event created little excitement. There was no fanfare, not even extensive press coverage—just one short paragraph in the newspaper. But the Wrights were not discouraged. The following year they built a more powerful airplane, and flew it from Huffman pasture, near Dayton. Then, in 1905, they built a third plane, the first fully practical airplane in history. With ease it could take off, bank, turn, circle, and do figure eights, and it could stay in the air thirty minutes at a time.

For several years thereafter the Wrights stayed in relative obscurity. Dogged by spies and bureaucratic obstruction, they kept themselves and their planes from public view. Meanwhile, Europe picked up the link of aviation progress. In 1903, before the Wrights made their first flights, Octave Chanute had traveled to Paris where he lectured on the brothers' gliding achievements. This sparked tremendous experimentation (much of it worthless) on the part of Europeans to achieve flight. Not until 1906 did the Brazilian, Alberto Santos-Dumont, in an ingenious but impractical biplane, make the short, twenty-second "hop" flights that gained him the dubious honor of innovating heavier-than-air flights in Europe. The following year Henri Farman made a one-minute flight in a biplane of his own design, and in 1908 Louis Bleriot, a prosperous manufacturer of auto lamps, built his famous monoplane, the *Canard,* which had paper-covered wings.

The following year, 1908, Bleriot flew the English Channel, and the Wrights emerged from their self-imposed obscurity. Deciding to fly in public for the first time, Orville toured the United States, while Wilbur dazzled Europeans during a six months' schedule of exhibition flights. At Le Mans, attired as always in his familiar cap and collar, Wilbur had spectators gasping as he banked and turned in a series of aerial stunts.

When the Wrights completed their respective tours, aviation still did not surge ahead in the United States. Cautious, the two brothers froze their monopoly. They built a few airplanes, which they sold at $20,000 each. They taught a few pilots to fly, charging high fees. They also leased a few of their machines for state fair exhibitions. Their policies slowed up the development of aviation in America. Highly religious, they leased planes, for example, only on the condition the planes not be flown on Sundays.

As a result, others, such as Glenn H. Curtiss, of Hammondsport, New York, began to enter the field of aircraft design and manufacture. An ex-motorcycle daredevil who had started manufacturing motorcycles, Curtiss was encouraged to enter aviation by a syndicate supported by Alexander Graham Bell, inventor of the telephone. For control, Curtiss's planes used not the warpable wings pioneered by the Wrights but ailerons—hinged movable parts of the airplanes' wings, located along the trailing edge. The Wrights sued for infringement of patents, but Curtiss went ahead anyway. Soon he was producing in small quantities a little single-seated biplane, powered by a 30 hp four-cylinder engine which he also built himself. In this and larger planes that followed, Curtiss flew in exhibitions, and in 1910 achieved national fame when, in a more powerful 60 hp flying machine, he flew nonstop from Albany to New York City. From Curtiss also came the world's first seaplane.

Soon improvements and innovations in aircraft design and manufacture were being made all over the world. However, with the exception of Curtiss and Glenn L. Martin, who in 1909 built his first plane, most aircraft development took place in Europe, where men like Anthony Fokker were beginning to build planes, and air heroes like the Frenchman, Adolphe Pegoud, and the Russian, Nesterov (each of whom holds the disputed claim of being the first to loop-the-loop), were becoming famous. From Europe at this time came monocoque (single shell) construction, and, ominously, the first fighter plane, the B.S.I, built at Farnborough, England. By 1914, the aviation industry in Europe had developed to the point that English, French, German, and Russian aircraft manufacturers were able to turn out the thousands of fighters and bombers needed in World War I. Names of their planes, like the Fokker monoplane, the war's first successful fighter; the Nieuport; Spad; Sopwith Camel; Handley-Paige; Bristol; Pfalz; Albatross; and Sikorsky—all were soon on everyone's lips.

American air power was virtually nonexistent when Congress declared war against Germany and her allies in 1917. All the army had were fifty-five training planes, and the navy's total complement was fifty-four training planes. So far had American aircraft design lagged that the United States did not have, even in blueprint stage, an airplane that could be used in battle. For war planes, the government humiliatingly had to copy English and French designs. Most of the planes manufactured in the United States were De Havilland DH-4s, a two-place observation craft produced under British license. This was the only United States machine to see active service. The Curtiss JN-4 "Jenny," also turned out in large quantities, was used for training only.

In the case of engines, however, the United States played a more important role. As with the planes, no acceptable engines were available at the beginning of the war. Curtiss had produced some 90 hp OX-4 engines, but the company's 200 hp engine had been unsatisfactory. When the United States began hostilities, a committee was organized among airplane and auto manufacturers to design a better engine. Two important members were J. G. Vincent of the Packard Motor Car Company of Detroit and Charles Hall of the Hall-Scott Motor Company of Los Angeles. Vincent had been working for several years with the German Mercedes company in trying to develop a design for an airplane engine, and he had studied the technique of welded cylinders. When he traveled to Washington, D.C., to work on the design with Hall, he had a good idea of the engine that would be needed, except for size. According to one account, the two men locked themselves in a hotel room with a drawing board for four days, and when they came out, they had the design for what was to become the famous Liberty engine. This engine, water-cooled and of a V-8 type, was 300 hp, giving 1,250 revolutions per minute. Its eight cylinders (later increased to twelve) were individually fitted and had welded steel jackets. Its valves were overhead.

Within three months, an actual engine was built, and, unbelievably, it delivered its horsepower in its very first trials. The nation's aircraft industry, with some help from the auto industry, mobilized into a cooperative group to produce the engine, and within a few months the four additional cylinders were added, bringing the engine's rating up to 400 hp. Production started soon after, and eventually reached 150 engines per day.

When the war ended, the United States sank back to its sad level of prewar aviation inactivity, leaving most progress again to be initiated by Europe. One factor facilitating the growth of passenger air travel in Europe was the availability of bombing planes. By virtue of their size and load capacity, they could easily be adapted to passenger-carrying and cargo-carrying planes. All that had to be done was to install seats and enclose the cockpit. But the United States, which had neither built nor used bombers in the war, had no big planes to convert. Most planes available here were similar to the small two-seater De Havilland DH-4s which could carry only one passenger (plus the pilot) and limited quantities of freight.

De Havilland DH-4s were used in the only large-scale aviation operation functioning in the United States outside the military. This was the airmail service, at that time operated by the United States Post Office itself.

During the war, the federal government had established airmail service. The initial objective was to launch operations between New York and Washington, D.C., and in early 1918, army pilots flying Curtiss Jennys laid out the route. Then, on May 15, 1918, with both De Havilland DH-4s and pilots borrowed from the Army Signal Corps, service was inaugurated. On hand in Washington to see the first airmail plane take off was a group of notables, including President Woodrow Wilson, Postmaster General A. S. Burleson, and Assistant Postmaster General Otto Praeger, who was in charge of the new service. Soon after, civilian pilots employed by the Post Office Department took over, and seventeen De Havilland DH-4s were shaken loose from the army. Rebuilt to handle mail at a cost of $2,000 per plane, these De Havillands became the nucleus of the new United States Air Mail Service, as it came to be called.

Although airmail service originally was restricted to the New York-Washington route, Praeger was determined to extend it across the nation. Despite the fact that there were no navigational aids and no emergency landing fields, he wanted to start up coast-to-coast service as soon as possible. In the months after the first flights, he gradually filled in some of the eastern and midwestern routes. In 1919, Chicago to Cleveland service was initiated; then followed the "hell stretch" over the Alleghenies from Cleveland to New York, and, finally, further extensions to Omaha and San Francisco.

The De Havilland DH-4, another World War I plane. By the mid-1920s, it was still being flown by the United States Air Mail Service.
AMERICAN AIRLINES

By September, 1920, Praeger had his transcontinental service. At night the airmail still had to be transported by rail, but because the mail was flown by plane in the daytime, Praeger was able to reduce the all-rail time of ninety hours across the continent to seventy-two hours. Unfortunately, neither the public nor the government appreciated the efforts that Praeger and the Air Mail Service were making. By 1921, most government officials considered the airmail experiment a failure. They saw little improvement in cutting off just eighteen hours from the rail time. They were appalled at the fact that of the forty pilots who had originally pioneered the coast-to-coast service, ten had lost their lives in crashes. The incoming Harding administration would be even less favorably disposed to the Air Mail Service. Not only had it put itself on record as opposed to most of the innovations, like airmail service, of Wilson's administration, it was also committed to a "return to normalcy," which meant reducing federal expenditures wherever possible.

Praeger realized that the fate of airmail service, and, indeed, of all civil aviation, was at stake. To save the operation, he and his courageous band of civilian pilots dreamed up a dramatic stunt: they would publicize the value of the Air Mail Service by doing something that had never been done before. They would fly the mail coast to coast entirely by air—flying by night as well as by day. Their challenge was almost insurmountable. There were no beacons, no landing lights, and no navigational aids for nighttime flying. For the dramatic transcontinental flights, bonfires would have to be lit on different airfields along the route.

Preparations were hastily made. The day chosen for the flights was February 22, Washington's Birthday, less than two weeks before Harding would be inaugurated and a new Congress would go into session. The plan called for two flights to start out from the west coast, and two from the east coast. On the fateful day, both westbound flights were washed out when the pilots ran into unfavorable weather on the graveyard run between New York and Cleveland, and one eastbound flight ended in a crash in Nevada.

The other eastbound flight got off successfully, the pilot in a De Havilland DH-4 Liberty-engined plane taking off from San Francisco's Crissey Field, climbing to 12,000 feet, clearing the Sierra Nevadas, and landing at Reno, where another pilot took over, flying the mail to Salt Lake City. From here, another exchange of pilots took place, the plane then being flown to North Platte, Nebraska, where pilot Jack Knight was waiting. Just out of a hospital from a broken nose and other injuries suffered the week before in a rough landing, Knight took over. Guided by bonfires set by volunteers at several Nebraska cities, he made his way to Omaha, where a tiny group of spectators waited at the chilly primitive airport. When they hear the sound of Knight's motor, they ignited drums of gasoline to light up the field, and Knight soon dropped down out of the sky.

But when he looked for his replacement, no pilot was there. The man supposed to have flown in from Chicago had not arrived. Knight had never traveled the Omaha-Chicago route, not even in the daytime, and he was already tired. But, determined that the transcontinental flight should not be canceled out, he gulped down some coffee and took off again. The winds were unfavorable and navigation, almost impossible. At Des Moines there was too much snow to land, so Knight flew on to Iowa City, an emergency stop. After gunning his motor, he saw no lights. With his fuel supply low, Knight started thinking of alternatives, when suddenly he saw a red flare. Everyone had gone home, but a watchman stationed there had heard his plane.

The next stop, and for Knight the final stop, was Chicago's Maywood Field. Day was breaking as his plane drew near, but a heavy fog covered the field. When it momentarily cleared, Knight headed down. His faithful Liberty engine was coughing and faltering, but he knew he could make it. A jubilant crowd was waiting for him. Word had spread by telephone that he was coming, and patrons of Chicago nightclubs, still dressed in evening clothes and dinner jackets, had driven out to the airport for the event.

Another De Havilland DH-4 was already warming up, and in this plane, pilot Ernie Allison picked up the mail and flew it to Hazelhurst, Long Island, New York. Praeger and his pilots had accomplished their goal. Transcontinental flying time had been cut in half—to thirty-three hours and twenty minutes. Knight's feat was lauded coast to coast. The new Congress, prodded by President Harding, quickly enacted laws calling for the continuation and expansion of the airmail service. An appropriation of $1,250,000 was made to light the air routes. And, most importantly, the first laws were enacted authorizing federal regulation of civil aviation.

With a new lease on life, the Air Mail Service set out to light its transcontinental route with beacons, but four years were to elapse before this lighting was completed. Meanwhile, the service's pilots kept risking their lives over the dangerous skies. Flying by instinct and intuition, they performed navigational miracles.

In the West, the tips of the Rockies served as navigational guides. But in the East, when fog covered the ground, a pilot flying above the fog had to fly blind. Some pilots judged direction and distance by the time it took them to smoke a certain number of cigars. Others carried telephone numbers of farmers along their route. Landing at one field, they would telephone ahead to see what the weather was like.

Not unexpectedly, morale among the pilots often became low, and Praeger's successor—Colonel E. H. Shaughnessy, who had come in with the new administration—initiated a series of moves to improve it. He ordered that pilots would have their own planes assigned to them, and they would not have to fly in any others. He also gave to the pilots, rather than to the Air Mail Service, the decision of whether to fly on a particularly bad day.

Shaughnessy was making headway when he died tragically in the collapse of a Washington theatre roof. Replacing him as head of the Air Mail Service was Colonel Paul Henderson, whose father-in-law was chairman of the all-important Ways and Means Committee in Congress. With a powerful voice in the legislative branch of the federal government, the Air Mail Service was able to accelerate its beacon program. Every twenty-five miles along the New York-San Francisco route, now known as the Columbia route, beacons were installed, and near each one an emergency landing field was also constructed. By mid-1924 almost two thousand miles of lighted airways had been completed, and that same year transcontinental airmail flights began, with planes traveling across the nation in thirty-two hours.

The airmail service was about the only bright spot in United States commercial aviation then. The big boom everyone had expected after World War I in passenger and freight airline operations had not materialized. The nation's financiers were wary of an industry that promised little or no return for years to come. No substantial businessman had yet invested in either airplane manufacture or airline operation. Commented the *New York Evening Post* in 1921: "Public interest in the use of airplanes for commerce and sport in this country has not yet reached the point where the industry can be supported on a commercial basis."

With aviation at such a low ebb, it was almost impossible for major innovations in airplane design to be made. Those bright innovators who, during the war, had begun to pick up where the Wrights, Glenn Curtiss, and Glenn L. Martin had left off were literally squelched by the lack of financing and support. Grover Loening, one of the new breed of airplane designers, commented later about that period: "Our great opportunity looked gone forever."

But some new pioneers never lost faith, and they managed not only to survive intact the aviation wasteland, but also to forge ahead with the fresh brilliant aeronautical concepts that by the end of the 1920s would return the United States to aviation supremacy. One of these men was William Bushnell Stout, an inventor-genius who—together with Henry and Edsel Ford—was responsible for the famous "Tin Goose," the plane that launched commercial passenger aviation in the United States.

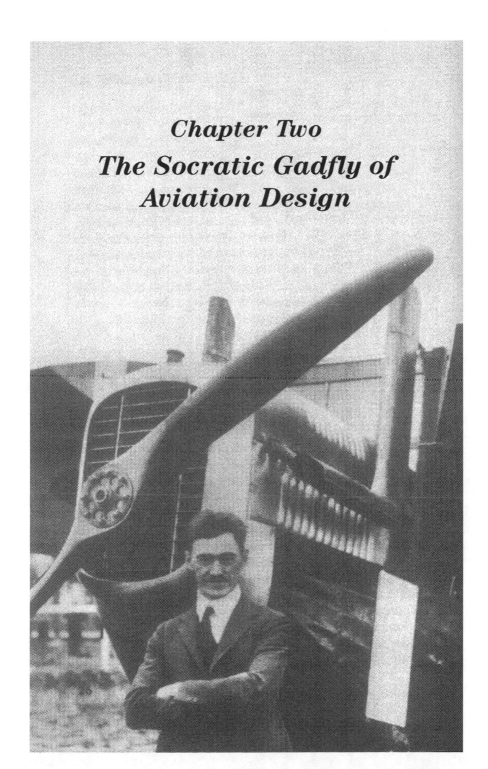

Chapter Two
The Socratic Gadfly of Aviation Design

By the early 1920s William B. Stout was already one of the nation's most brilliant and unpredictable inventors. To anyone seeing Stout for the first time, he actually looked like an inventor, or, rather, like most people's idea of one. Tall, skinny, sometimes moustached but always bespectacled, he usually wore tweed suits, and, more often than not, sported colorful striped shirts and gaily-dotted bow ties. His most unforgettable physical feature was a wild mop of thick bushy hair that he pushed around when he got excited.

Stout also worked as hard as inventors were supposed to. On many a project he labored around the clock until he obtained the results he sought. And he obviously had ability. As all those who worked with him could testify, Stout could turn out new mechanical contraptions the way most people make doodles. By the early 1920s he had already supposedly patented hundreds of inventions, and would eventually have more technical patents than any American except Edison.

Unlike most inventors, Stout also possessed an unusual gift for being able to promote his inventions successfully. As historian Allan Nevins once said, Stout had "a genius for clarifying complex ideas," and he could do this in many media. He was an accomplished pen-and-ink artist. He was an articulate speaker who could illustrate a technical point with a happy vein of humor. He was also an excellent writer, able to come up with the telling phrases that could get himself and his ideas across.

In newspaper interviews, Stout admitted frankly that his business was dealing with "screwball" ideas, and he emphasized that he was proud of this, since screwball ideas were, in his opinion, responsible for

Opposite: William Bushnell Stout, who conceived the idea of an all-metal commercial airplane being manufactured in the United States and who formulated the principles that led to the Ford Tri-Motor. FORD ARCHIVES PHOTO

the world's progress. He also coined quotable mottoes like "Laziness is the mother of invention." And he explained his working philosophy by saying that he wanted to know not only *why* wheels went around and *how* one could make them go around, but also *why* anyone would *want* to make them go around. In tackling such questions, Stout often found that the accepted answers were based on false assumptions, and when he went beyond these, he usually achieved his best inventions (and the most publicity).

Both before and after the 1920s, Stout worked in many other fields besides aviation, but it was in aviation he did his best and most famous work. Here his peculiar personality and approach made him, as Owen Bombard once so aptly put it, the "Socratic gadfly of aviation design," and yet his gad flying produced results. If not for William Bushnell Stout, there would never have been a Ford Tri-Motor.

Stout's story starts in the post-Civil War era. America was on the eve of great expansion. The bloody War Between the States had ended only fifteen years before, and the nation was still operating largely on a regional basis. Industry was just developing on a large scale, and men worked hard—fourteen hours a day, six days a week—for incredibly low wages. Agriculture was still supreme. Eighty-five of every hundred men in the nation's labor force were engaged in growing food. As for communication and transportation, both were in a primitive stage. Mail was slow and uncertain. There was no national advertising. Newspapers circulated in local areas only. In transportation, except for the railroads, everything depended on horses. There was no transcontinental road; there were few paved streets. Freight rates were high. With horses able to make only fifteen miles per day, it cost eighty-five dollars to ship a barrel of flour from New York to Boston via horse-drawn vehicle.

But signs were appearing to indicate that the situation would soon change rapidly. The railroads were greatly extending their steel rails. The telegraph was being used more widely. New inventions like the camera, the sewing machine, the reaper, and the typewriter foretold the approach of a new America.

In the middle of this era, on March 16, 1880, William Bushnell Stout was born in Quincy, Illinois. Stout's father, the Reverend James Frank Stout, was a Methodist minister. He was descended from a Captain Stout, an Englishman who had settled in Virginia in Colonial days. The Reverend Stout's father and mother, Stout's grandparents, had emigrated to Illinois from New York State in a covered wagon. A preacher too, Grandfather Stout also had a trade of mending shoes. In Illinois he became a preacher-cobbler, caring (as Stout later humorously described it) for men's soles on weekdays and men's souls on Sundays.

Stout's mother was Mary C. Bushnell. Her family line also went back to Colonial days, and one of her collateral ancestors was a famous inventor—David Bushnell, who during the American Revolution built a primitive submarine (The American Turtle) and torpedo with which he unsuccessfully attacked the British. Her father, Stout's other grandfather, was a well-known Chicago contractor, who, among other projects, had constructed the Grosse Pointe lighthouse in Evanston, Illinois.

Stout's father was a disciplinarian, but he was also kind, patient, and tolerant. From him the entire family received inspiration, and Stout all his life never forgot some of the things his father told him, such as "Learn the most things and you will live in the most worlds."

When Stout was five years old, his father took him to a beach that fronted on a lake. Here they stood, hand in hand, watching the gulls sweeping along the water, their wings driving silently, raucous calls issuing from their throats. Stout began asking his father questions about flight. "Some day men will fly like that, just as soon as they know how," his father said. "And can I?" the boy asked. "You might," the father replied. "You might even become the first man to do it."

Stout never forgot that moment, for somehow it gave him, as he later put it, "purpose." No man had ever flown like the birds. Maybe he would be the first. In any case, from that time on, he began to rapidly pick up various skills. He began to draw. He learned to play the piano by ear. Later, he even taught himself foreign languages, using his own method of concentrating on word lists instead of grammar.

During these early years Stout's father was relocated many times; and in 1890, when the family was in Minneapolis, a teacher discovered that Stout's eyesight was so bad he could not see the blackboard. An oculist prescribed thick-lensed glasses, and from that time to the end of his life Stout was never without eyeglasses. The immediate effect of these new spectacles was that Stout's interests became different from those of his classmates. They spent their extracurricular activities in pursuits like playing ball; Stout spent most of his spare time at home. He had always been interested in toys, and now with time on his hands—and family income too low to purchase toys—he began to make them himself out of odds and ends around the house. Soon Stout was so proficient at devising and whittling toys that he began to teach his friends how to make them.

About this time the family took its only pleasure trip within Stout's memory. Grandfather Bushnell invited them all to Chicago during the famous World's Fair of 1893. Stout was thrilled. It was his first visit to a large metropolis, but most of all, he was excited by what he saw at the fair—the vast installations of electric lights and the mechanical wonders in Machinery Hall.

Shortly after, when the family moved to Mankato, Minnesota, Stout began to get more and more involved with machines, and he became interested again in flying. *Youth's Companion,* a leading children's magazine then, published an article that explained how to build a model airplane out of a cork, two feathers, two rubber bands, and tissue paper. Stout followed the directions, and his plane was so successful it flew to the top of his father's church.

In 1897 his father was transferred again, to Saint Paul, Minnesota, and by now, the mechanical abilities of Stout at seventeen were so recognized that he was accepted at Saint Paul's Mechanical Arts High School, the first manual training school in the west. Being extremely handy with tools, he built himself not only a cottage at White Bear Lake that summer, but also all the cottage's furniture. Before young Stout finished high school, his father was relocated again, but Stout, in order to complete his courses, stayed on in Saint Paul, rooming and boarding with family friends. To help pay expenses, he got a job copying papers for a patent attorney—gaining experience in the details of patent procedures that later would prove invaluable.

After graduating from Mechanical Arts High School, Stout attended Methodist University in Saint Paul at his father's suggestion, but after two years he quit to teach in a small country school outside Winona, where his family was now located. It was a one-room schoolhouse with eight grades and thirty pupils, and to help teach arithmetic, Stout invented what may have been the first teaching machine. It was a clocklike gadget with hands, controlled by rubber bands, pointing to numbers around a dial. When the class was on, say, "sixes," a pupil in the front of the class would pull a string that would move the hands to some number like "four." The first pupil in the class to call out "twenty-four" would get a mark. So many marks gave a student the privilege to pull the string next. By the time Stout had the machine in operation a month, even first-graders knew their multiplication tables backward and forward.

Stout now was making other machines in a shop he fixed up behind his home. One of these was a homemade phonograph constructed of sewing spools, a wooden pillbox, and a piece of paper used as a diaphragm. For a motor (since he had no electric motor) Stout built a water engine that he fastened to a hydrant in the yard and connected to the phonograph by means of a long string belt. By the hour he ran his phonograph, playing only one record, since he could not afford to buy another.

After teaching two years at Winona, Stout decided to go back to school, and he enrolled in the engineering college of the University of Minnesota. Money was still a problem, so he took on many odd jobs and projects. He organized a laundry service for students. He formed a collegiate shopping-discount association by which local firms—paying $50 to participate—gave discounts to students. He also for the first time began to capitalize on his toy-making abilities. Remembering a toy he had built in his Winona workshop, Stout wrote up the directions on how to make it, and sent them off, complete with drawings, to the "Round

Table" column of *Harper's Magazine.* A week later, he received a check for twelve dollars.

From then on, Stout was a professional writer on toy-making, and eventually he was able to pay almost half his college expenses with articles on toys. Magazine after magazine—*Harper's, Youth's, Companion, Scientific American,* even an anti-cigarette publication—published his sketches and how-to-build-toys articles.

The secret of Stout's success was that he realized that most boys had little money to invest in toy components. Thus, the toys he wrote up could be put together from inexpensive objects found around the house, like string, spools, paper, pieces of wood, and nails. Also, Stout's toys were simple-to construct, and his directions, easy to follow.

Some of Stout's toy ideas he was able to transform into practical devices. The following year at the university he arranged to get free rent by tending furnace. His room was on the third floor of a home owned by an elderly lady who doted on comforts, one of which was a firm, unchangeable furnace schedule. At 4 A.M., his nightshirt flapping in the breeze, Stout would run down three flights to the basement to drop the furnace's damper and raise the draft, so the house would be warm by the time the lady woke up several hours later.

After several weeks of studying late and rising early, Stout got an idea. Why not run strings from the basement furnace to the window outside his third-floor room? Then, at 4 A.M., when his alarm went off, all he had to do was open the window, unhook one string to drop the damper and the other to raise the draft, all without going near the basement. The mechanism worked fine for several weeks until one of the strings broke. Then Stout invented an improved device. To his alarm clock keys, he tied strings that connected with crossbars attached to the furnace's damper and draft controls. Now he had an automatic furnace tender. When the alarm clock went off, the mechanisms began operating without his even having to get out of bed.

Shortly afterward, Stout was asked to write up a handyman's column in the *St. Paul Dispatch.* Describing the automatic furnace tender, Stout got his usual fee for the column—five dollars—but three months later, he read where a new company was formed in Minneapolis to manufacture heat regulators using alarm clock devices. Stout never found out, but he often wondered how much debt this company (which later became Minneapolis-Honeywell) owed to him and his column.

Stout completed his studies at the University of Minnesota until the final examination week of his last year, when he was struck down by a corneal abscess. An oculist warned that he would lose his sight if he took his finals. It was a bitter blow, but that summer, when his eye grew better again, Stout followed through with plans he had made the previous winter to travel to Europe. He had written many articles in advance to accumulate funds for the trip and had arranged for overseas writing assignments. Somehow he managed financially, visiting France, Belgium, Switzerland, Germany, and England, and working his way back on a British freighter.

The trip was valuable in many ways for Stout. First of all, it taught him a lesson he applied the rest of his life, that a person can do anything if he wants to enough and is willing to pay the price. Secondly, it renewed his interest in aviation. While Stout was in Europe, he heard the Europeans talk of little else but aviation. Santos-Dumont, Octave Chanute, the Wright brothers—everywhere Stout went, he heard about flying and aviation.

As a result, after his return Stout began experimenting with glider models. When he had first begun writing about toys, he had built many gliders. Cutting pine sticks the size of lead pencils for wing and plane frames, he would tie them together (there was no good glue then) in the configurations he wanted. For the wings, he would use tissue paper which he put on damp; when it dried, the wing surfaces were tight as a drum. In those days, all wing surfaces were flat (curved surfaces in models were as yet nonexistent), and balance had to be achieved by proper positioning of weight. But Stout's models flew, first in random dives and flops, later in long glides as he mastered the design and came to know what really made a plane fly.

Becoming more interested in aviation, Stout arranged a meeting with Octave Chanute in Chicago. Shortly after, Chanute asked Stout to address the Saint Paul Engineering Society on "Artificial Flight." Chanute himself had been invited to speak, but because of ill health, was unable to make the trip. The French-born engineer had suggested Stout as a substitute and sent Stout his slides. The speech went over well, and Stout's interest in aviation became even more intensified.

Meanwhile Stout had married, and to support himself and his new wife he took a job teaching manual training at Saint Paul's Central High School. He was still earning extra money with his toy and manual arts columns, and no longer were they restricted to the *St. Paul Dispatch.* The McClure Syndicate had picked them up and was now distributing them all over the nation. As in his previous columns, Stout stressed using odds and ends around the house for building the toys. The only tools needed, he told his readers, were a file and a jackknife. He described how to drill with the file's sharp handle end, and mentioned some of the million and one uses of the jackknife. From this latter tool he took a pen name that he soon signed to all his columns—Jack Kneiff.

Shortly after this, Stout was able to give up his high-school teaching job and work full time at his column. So profitable was his writing now that he was able to travel again to Europe, this time with his new bride. While there he wrote articles on aviation, which he illustrated with pen drawings.

Upon his return to the United States in 1910, he learned that a group of promoters was organizing an air show at the state fairgrounds in Saint Paul. Everyone in the Twin Cities was excited about the forthcoming event. Wanting to participate in some way, Stout arranged to stage as one of the show's events a model-airplane meet where rubber-band flying models would compete for prizes.

The air show was a tremendous success. The model-airplane meet went off as scheduled, and so did the flights of the real airplanes. One of the flyers had never flown before. He was an inventor who showed up in a plane of his own design. Since it had neither ailerons nor wing warping, control was impossible. As soon as he got into the air, he would begin having difficulty and would eventually hit the ground, spinning around and splintering a wing. This man made flight after flight, putting on a new wing each time. Not long after this meet, the man—who was Lincoln

Beachey—became the greatest stunt pilot of his day and America's greatest aerial daredevil rival of the Frenchman, Pegoud.

Another flyer that day was Eugene Ely, later to become the first pilot to take off and land from an aircraft carrier. Stout worked as a mechanic on Ely's plane, and, like Beachey's, it too lacked complete control. The trouble here was insufficient power, which prevented Ely from turning. Every time Ely tried to turn, he was forced down.

On the last day of the air meet, Glenn Curtiss himself showed up in his special 60 hp "June Bug" plane. Making his last public appearance as a flyer, Curtiss flew around the fairgrounds at 200 feet without coming down for what seemed like hours. The audience cheered, and left the fair talking of nothing but aviation.

Glenn H. Curtiss at the wheel of his 60 hp "June Bug" plane, PICTURE COLLECTION, NEW YORK PUBLIC LIBRARY

The Curtiss "June Bug," equipped with pontoons, being tested on an upstate New York lake in 1908. PICTURE COLLECTION, NEW YORK PUBLIC LIBRARY

After the meet, Stout wrote more articles on aviation, and his writing became more technically oriented. In 1912, the *Chicago Tribune,* about to start an aviation column, invited him to become its aviation editor. Stout took the job on the condition he could still write his Jack Kneiff columns, and he moved to Chicago. As in the Twin Cities, he organized model-airplane clubs among high-school students. Some of these students later became prominent in the aviation industry, among them James McDonnell, founder of the giant McDonnell Aircraft Company, now McDonnell-Douglas.

As Stout continued his writing for the *Chicago Tribune,* he was also made technical editor, and this led him into other fields like that of automobiles. Years before, he had built a gasoline-driven vehicle which had looked something like a baby buggy, but it had been accidentally wrecked by a friend. Later, after his second European trip, Stout had constructed an improved kind of motorcycle to which he had added certain automobile-type features such as an auto seat and springs. But driving his Bicar (the name he gave it) one day in Duluth, he ran it at top speed and flattened the ball bearings, ruining the engine, and temporarily stopping his experimentation in that field. In 1914, knowing of Stout and the Bicar, the Scripps-Booth Motor Company of Detroit hired him away from the *Tribune* to help design a new line of motorcycles and automobiles that the company intended to manufacture. Stout started out as the company's chief engineer, but soon was also put in charge of Scripps-Booth's advertising and sales.

Up to this time Stout had never even driven a real automobile, but that did not stop him from making an innovation in automotive design. As Stout saw it, the public wanted a car that would not only be comfortable but would also have economy of space. Explaining this one day to Scripps-Booth executives as he was walking through the factory with them, Stout picked up a piece of chalk. On the floor he began sketching his concept of a new-type single-seat, topless car he felt the company should produce. Scripps-Booth did manufacture the car—picking up the design from the chalk marks Stout had sketched on the floor. According to Stout, it was the first roadster in American automotive history.

Despite the demands of his Scripps-Booth job, Stout continued to write his Jack Kneiff columns, and he also kept designing toys. Many of these he was able to sell on a royalty basis to toy manufacturing companies, and from the profits he received he financed his experiments with flying-machine models. Around his Detroit home in the early evenings and holidays, airplane models were always flying.

Stout's working with airplane models led him to two concepts new to aviation at the time: one, that current aeronautical design was going in the wrong direction—that from a design point of view, thick-wing monoplanes were superior aerodynamically to the thin-wing biplanes then being manufactured; and two, that aviation had a great commercial future with airplanes flying payloads of passengers and freight—a concept differing from the prevailing attitude that the airplane was useful only as a military weapon.

Unfortunately, the advent of World War I reemphasized the military utility of airplanes, and even Stout himself entered the military-plane field. When hostilities broke out, the United States boasted it would soon be producing 25,000 planes a year. But this was a hollow boast, since the aviation industry was too embryonic to support such a manufacturing effort.

In 1917 Alvan Macauley, head of the Packard Motor Car Company, called Stout in. Through J. G. Vincent, the Packard executive who had co-designed the Liberty engine, Packard was getting ready to manufacture both the engines themselves and the planes they would be installed in. To accomplish all this, Packard was organizing an aircraft division, and Stout was offered the position of chief engineer.

Stout accepted the new job, but hardly had he started when Packard sent him to Washington, D.C., to serve as technical adviser to the newly created Aircraft Board under the chairmanship of Howard A. Coffin, a well-known automotive company executive.

Stout moved to the nation's capital with his family (he now had a daughter), and while he worked at expediting the manufacture of war planes, he also began to realistically assess what was right and what was wrong about the airplanes of the day. As Stout began to see it, the current state of aeronautical design left almost everything to be desired. The planes then available, in both Europe and the United States, had no real speed. They had no real brakes (just a tail skid in the rear). There was no way to steer them on the ground except by their rudders. In Stout's opinion, these and other deficiencies made aviation far too dangerous and difficult. Further, he believed that there were just two basics in airplane design: (1) structure, i.e., what holds the plane together in flight, and (2) aerodynamics, i.e., the plane's performance in regard to lift, air flow, and control. Stout could see that the aerodynamics of a plane could be compromised to some extent; even a few mistakes could be made. But this was not so with a plane's structure. What good was aerodynamic performance if the plane's wings fell off?

The trouble was that no one really knew what was needed technically. When the first 400 hp Liberty engines appeared, one was installed on an old Curtiss biplane previously powered with a 200 hp engine. With double its previous power, the plane flew only 5 mph faster. Stout reasoned that an unknown factor was entering the picture, a factor not important in land or sea vehicles. This he came to realize was *drag*. After the War Department set up a wind tunnel at McCook Field near Dayton, more information became available on this vital subject. It was found, for example, that a small wire, vibrating crosswise from a wind sweeping over it, had more drag than a solid strut 1 1/2" in diameter. From then on, planes began to be designed with diagonal wing struts. But Stout felt designers should go farther. He calculated from analyzing wind-tunnel statistics that when a standard biplane with a 400 hp engine was involved, 297 of the horsepower it delivered went into shoving through the air the plane's struts, braces, radiator, and landing gear, leaving only 100 horsepower for the plane's actual lift.

The key to obtaining more power, Stout reasoned, was monoplane design. He saw an ideal plane as one that would be in effect nothing more than a large wing, with control surfaces and even the engine hidden inside. Why not a monoplane? Stout argued. After all, would Nature favor a monoplane design for birds if it were not better?

The more Stout thought about it, the better his idea of a thick-winged monoplane seemed. He decided to design such a plane. This plane, which came to be called the Batwing plane, and also the Vampire Bat, had such an unorthodox design that those of the Aircraft Board who saw it got what one observer called "technical colic." With cantilever construction, it had no exposed struts or wires, and was entirely internally braced. Essentially the Batwing was a "flying wing," a design that aeronautical engineers would not come around to for three more decades.

Such a plane had never been built before, in either the United States or Europe. But Stout was not daunted. "If the wing of an airplane is the only part that lifts, why should anything else but the wing be exposed?" he argued. Working up a brochure explaining his new revolutionary-design plane, he got his point across about drag by comparing a plane with exposed struts to a motor boat slowed down because it was towing a load. Remove the load and the boat goes at full speed, Stout said. Remove the outside struts and wires from a plane by bracing it internally, and a speed of 210 mph instead of 110 mph can be attained.

The Aircraft Board was flabbergasted at Stout's unusual Batwing design, but was not unsympathetic. With the war coming to an end, the board was exploring new concepts. Other equally revolutionary concepts that Stout had nothing to do with were also being proposed to the board—metal propellers, variable-pitch propellers, air-cooled radial engines.

As a result, the board gave Stout the go-ahead to construct a trial model of the Batwing. Stout quickly made the model, and the board received a shock when Orville Wright, called in to examine it, called it "the next step in aircraft." Stout was given a contract to produce a mock-up of the plane. He moved to Detroit to supervise construction, and, typically he decided that instead of building a mock-up, he would skip that stage and construct first an actual flying model.

As the design had been revolutionary, so was the actual construction. The Batwing was built of wood as other planes at the time were, but instead of fabric covering the wings, Stout—for a first in American aviation history—covered them with veneer. Making this possible was a new, recently developed waterproof casein glue plus the technical know-how of Widman Brothers, an expert Detroit automobile-body manufacturing firm, which Stout had engaged for the project. The wood used for the Batwing was spruce sticks and one-inch plywood board. The veneer, 1/16 of an inch thick, had a mahogany core with birch on the outside, all glued together under pressure with the new glue.

When the Batwing was completed, it looked exactly like its original model, i.e., a thick wing, shaped like an inverted Greek letter Delta and extending back to the tail surface. As in the model, all bracing was internal. Stout had succeeded in building the first internally braced cantilevered airplane in the United States, and in becoming the first to use veneer for an airplane's skin.

The Batwing still needed an engine, and this was supplied by Charlie Nash, the automobile manufacturer, who donated a 150 hp Hispano-Suiza engine. Once completed, the Batwing was taken to McCook Field for tests, and the revolutionary plane performed successfully. But disappointment then followed. By the time of the tests, World War I had ended. Both the board and the government rapidly lost interest in aviation. Although Stout was paid for building the Batwing, the government gave him no more encouragement and withdrew its support. To continue his aeronautical experiments, he would have to proceed on his own.

Stout went to work and within a few months organized Stout Engineering Laboratories. His chief backer was Robert Stranahan, president of the Champion Spark Plug Company. Convinced that Stout's aeronautical theories were practical, Stranahan invested $15,000 to finance the design and construction of a cabin plane based on the Batwing design.

David A. Weiss

Stout's revolutionary Batwing planes. In the top photo is the Batwing cabin or Vampire Bat plane. The ten men testing the wing demonstrated Stout's publicity idea for showing the strength of the internally braced cantilevered wing. In the center photo can be seen the Batwing's retractable radiator, a type of streamlining unusual for the day. In the bottom photo is Stout's all-metal torpedo plane built for the navy. Also called "The Bat," it crashed shortly after this picture was taken, FORD ARCHIVES PHOTO

Stout, as with the original Batwing, continued his unorthodox approach. He had assistants now, and he kept telling them that real creative design did not follow tradition. In the case of aeronautical engineering, Stout insisted the objective should be eliminating parts, not adding them. "I get it," one of his men said. "You mean your goal is to simplicate and add more lightness." Stout nodded. The next day he had a sign made with that slogan, and for the next three decades he kept it on his desk.

Many Detroit automobile manufacturers dropped into the small factory to see how the new Batwing cabin plane was progressing. Among the visitors were Roy Chapin, head of Hudson Motors; Walter P. Chrysler of the Chrysler Company; and William Benson Mayo, chief engineer of the mighty Ford Motor Company.

In building the new Batwing cabin plane, Stout innovated again: making his own plywood. When completed, the plane had a configuration similar to that of the previous Batwing, but its structure was stronger. So strong was its thirty-six-foot-long cantilevered wing that in publicity photos Stout posed ten men standing on it.

Like the original Batwing, this plane was completely trussed internally and it had a 150 hp engine. It also had several new features, including engine radiators which could be retracted in and out of the wings to adjust for constant cooling at varying air temperatures.

The Batwing cabin plane was flown the first time in late 1921 at Packard Field in Detroit. The test pilot was Bert Acosta, who had once said, "Give me enough horsepower and I'll fly a barn door." Before a group of Stout Engineering Laboratories employees, Acosta warmed up the engine, swung the plane in line, and opened up the engine. Off the ground the Batwing rose, up over utility wires at one end of the field, after which it headed toward the city. A few minutes later, Acosta turned back to land. Stout, watching with others, suddenly noticed that one of the plane's two wheels had fallen off. Expecting the worst, he tried to signal Acosta, but the pilot did not seem to see him. It made no difference. Acosta set the plane down on one wheel, and held it upright until its wings lost their lift. The plane then slid over on its back. Fortunately the damage was only slight, and Acosta climbed out unharmed.

The United States Navy became interested in Stout's new plane, and requested that he make a special series of tests. On December 23, 1921, the Batwing cabin plane took off again, this time with 1,170 pounds of load. The test was extremely successful; in ten minutes the plane climbed to 5,000 feet. So impressed was the navy, it told Stout it would be interested in having him build planes for the navy.

Stout was already thinking ahead. Despite the success of the Batwing cabin plane, he saw certain faults in the design and in the designs of other planes of the day. There was the matter of pilot visibility, for example. No plane so far, not even his, afforded pilots really good visibility. There was an even more serious fault in airplane design, Stout now realized. This was the use of wood as the chief construction material. Wood might be satisfactory for those old biplanes with exterior struts and braces, but when used in internal trussing, too much strain developed on the wing tips, which sometimes cracked.

The more Stout thought about wood, the less desirable it seemed. True, it was light, strong, and inexpensive, but it had many disadvantages. It was difficult to obtain in the proper quality and grade. It often deteriorated when exposed to the atmosphere, particularly in airplanes where extremes in temperature and humidity were encountered. And it was almost impossible to tell when wood had deteriorated. So deceiving was wood's outward appearance, the only sure way to determine its strength was to break it open—an impractical method in the case of an airplane in service.

For other reasons, too, Stout found wood unsatisfactory. The very fact that glue and wood screws had to be used with it produced areas of weakness. Also, in case of an accident, wood splintered and broke up. And it had no inherent resistance to fire.

It became increasingly obvious to Stout that planes should be constructed not of wood, but of metal. And he was not the only aeronautical designer thinking in terms of metal as an aircraft construction material. In

Germany, Dr. Hugo Junkers had come to the same conclusion, and was already beginning to build all-metal planes.

In the United States this concept seemed ridiculous to just about everyone in aviation except Stout. However, he was not intimidated. Over and over he argued the advantages of metal: how it was stronger than wood, and more reliable; how its strength could be estimated within 2 percent; how it did not splinter on impact, but had a tendency to retain its shape; how it did not burn like wood.

Stout kept pushing metal as an aircraft construction material against industry opposition. While building the Batwing cabin plane, he had heard of a new aluminum alloy called duralumin, which the Aluminum Company of America had developed. Composed of copper and aluminum, duralumin was nearly as light as ordinary aluminum, but it had almost twice the tensile strength (55,000 pounds per square inch compared to 33,000 pounds per square inch).

After getting in touch with Dr. Earl Blough, the Aluminum Company's technical director, Stout became convinced that the new alloy was what he was looking for. His next plane would be a duralumin plane, Stout informed the industry, and he talked the navy into sponsoring not just one all-metal duralumin plane, but three. The arrangements were worked out over a period of months. The planes would be all-metal Batwing torpedo planes. Each would have two engines and carry one torpedo. The navy would pay Stout $50,000 for each plane.

Stranahan came forward with additional financing, and Stout rented a bigger plant in which to start construction. The new plant was on Beaubien Street in Detroit. Across the street was one of America's most famous factories—here Henry Ford had set up his first automobile production line.

At the new plant Stout assembled a group of young engineers, some of whom would work with him for years. He deliberately chose engineers inexperienced in aviation design, reasoning that they would thus be unhampered by industry conventions. His chief assistant was George Prudden, whom he had first met years before when Prudden and his brother came to Stout's house with a model airplane the brother had built. A recent University of Minnesota graduate, Prudden was a specialist in reinforced concrete structures. Stout gave him the job of analyzing stresses and of developing new structures for the new navy Batwing. Another assistant was Stanley Knauss, a young auto salesman whom Stout had met in his Scripps-Booth days. Knauss was made the new company's sales manager, and the business manager was Glenn Hoppin, an electrical engineer.

Stout Engineering Laboratories had many design, engineering, and technical problems to solve to build the navy Batwing plane. More pilot visibility had to be added. A new fuel system had to be devised. A powerful engine was needed. But the major problem was to develop the techniques for fabricating the aluminum alloy onto the structure of the plane. Riveting, fatigue resistance, corrosion—those were just a few of the factors that had to be investigated and evaluated.

Working closely with the Aluminum Company, the Stout group moved ahead. They designed the plane so most of it could be constructed with bent forms. The duralumin sections were rolled by hand over chilled cast-iron rollers. Months were spent developing a satisfactory framework spar. Since so little was known about heat-treating duralumin, Stout had to purchase the metal already heat-treated and work it cold.

Navy officials visited the plant frequently, and on one occasion examined the nose section which had been fabricated from a curved piece of metal. Since this section would take so much of the structural load and needed the most reinforcement, they suggested it be tested before construction proceeded. Production was stopped for a week while Stout designed a test jig which he installed on a big girder that extended to the top of the plant twenty-five feet above. The nose piece was mounted at the bottom, and more than fourteen hundred pounds of sandbags were placed on the girder, pressing down on the nose. Much to the navy officials' surprise, the nose was unaffected.

In 1922, the first of the navy's all-metal torpedo planes was completed. Designated the ST All-Metal Torpedo Plane, it was the first all-metal airplane ever built in the United States. As with all the Stout planes, the wing area dwarfed the rest of the plane, but unlike Stout's previous planes, this craft had two engines.

For tests, the ST All-Metal Torpedo Plane was hauled to the army's Selfridge Field, outside Detroit, and here a full-sized navy torpedo was installed under the fuselage. Eddie Stinson, the well-known Detroit test pilot, was hired to fly the plane, and he took it up successfully. Stout himself was a passenger on one of the test flights.

Shortly afterward came the navy's official acceptance tests. A delegation of admirals and Navy Department officials journeyed to Selfridge Field. Stinson again took up the ST All-Metal Torpedo Plane, and again the tests were successful. But the navy then insisted that one of its own pilots fly the plane. Obviously very nervous, the pilot climbed in and took the plane up. Circling the field once, he then started down, but barely cleared some nearby trees. When he reached the landing field, the plane was so out of control that he pancaked it into the ground. The aircraft's landing gear gave way and its wings folded in a cloud of dust. The navy pilot climbed out unhurt, but two years' work lay in a heap of ruins. Dejectedly, Stout returned to the plant. "It was the bluest day of my career," he said later.

The navy withdrew its support and canceled out the other two planes. All they ever paid Stout was $65,000, and the Stout Engineering Laboratories ended up about $150,000 in the hole.

With Stranahan unable to provide any more financing, Stout was down, but not out. Even though the future of his company was uncertain, he had no intention of giving up. Rather, he made a decision to switch course in his manufacturing. Instead of designing and building planes for the military, he would construct them for commercial buyers. He would build a plane as yet unknown to American commercial aviation. He would build an all-metal transport plane.

Chapter Three
The Father of Assembly-Line Production Tales a Flyer

Stout's decision to build an all-metal transport plane did not find him starting from scratch. Characteristically, he already had in his mind the general outline of such a plane. His "Air Sedan"—the name he had already thought up for it—he visualized as being large enough to seat up to four passengers and to carry freight too. He also saw its construction as being more advanced than that of either of the Batwings. While building the navy Batwing, Stout had been called in by Eddie Stinson to help repair a Junkers JL-6 plane. Imported from Germany, it had been damaged on one of its first American flights.

Stout saw on this plane, which was constructed of metal, something he had never seen before: corrugated metal, and immediately he realized its advantage. If an entire fuselage was made of corrugated metal, far more strength could be obtained than with the non-corrugated metal skin he had been using.

Before Stout could even think of building his Air Sedan, he had the more immediate problem of keeping Stout Engineering Laboratories in operation. Because of the financial pinch, the plant was almost shut down, and only with difficulty could Stout pay his bills. Somehow he always managed to pull through, and often his revenue came from unexpected sources. Royalty checks were still being received from toys he had invented in his Jack Kneiff days, and a new toy, which he had worked up as something of an afterthought, unexpectedly paid off.

Stout had devised it while constructing the navy Batwing. He had attended a Society of Automotive Engineers convention in French Lick, Indiana, and, although he did not play golf, he was persuaded to try it. As a result Stout learned the principles of driving and putting

Opposite: Henry Ford and Edsel Ford, who backed Stout with the resources of their mighty automobile empire, FORD MOTOR COMPANY

and he became intrigued by the possibilities of a toy miniature golfer whose club could be swung to hit a small celluloid ball toward a cup. Constructing such a device, Stout sold the manufacturing rights to a toy company. The arrangement was that he would receive royalties, but so engrossed was he in the trials of keeping his company alive, he forgot all about the toy golfer.

One day when Stout faced a pressing $2,000 bill, it looked as if he might have to default. Never were the fortunes of Stout Engineering Laboratories so low. And then, in the mail, came a check for $2,000 from the toy company—royalties on the miniature golfer.

Stout soon realized he could never finance the manufacture of his Air Sedan on such income as toy royalties. He consulted Stranahan, the sparkplug company executive, who advised Stout to reorganize and start up with a new group. "You need a number of very wealthy backers," Stranahan said. "Nothing else will work. You can't sell stock to the public, and with no collateral you could never get a bank to finance you."

Stout took the advice, and on November 6, 1922, he reorganized his company as the Stout Metal Airplane Company.

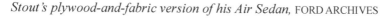

Stout's plywood-and-fabric version of his Air Sedan, FORD ARCHIVES

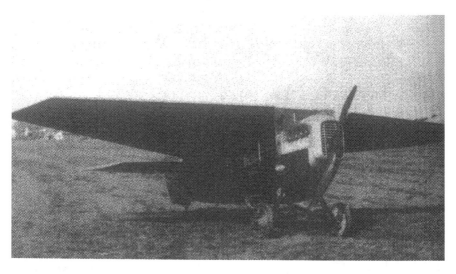

Then he created a financing campaign to this day unprecedented in American business in method and result. It consisted of little more than a series of letters directed to prominent businessmen, but, according to *Printers' Ink,* it produced more dollar results than any other American letter-writing campaign, and accordingly it ranks with the most successful promotions of all time. Equipped with no more than a typewriter and a total expense budget of less than one hundred dollars, Stout within a few months raised $128,000, and also obtained the backing of the two men in the world who could do him the most good—Henry and Edsel Ford.

Stout's sales letters were actually a correspondence course in aeronautics. The original list to whom he mailed his letters consisted of one hundred Detroit industrialists, mostly automobile manufacturers. In his first letter, Stout announced he was offering a weekly newsletter in which he would explain the fundamentals of aviation. These first letters Stout typed himself, and for them he drew original illustrations in the style of his old Jack Kneiff columns. At night his wife and his daughter, sitting around the dining room table, traced the illustrations onto the letters with carbon paper.

The text of the first letter read: "If a man walked into your office with a design of a motor car with its engine upside down and under the rear seat, you would quickly be able to answer him as to its value, and you would have the authority to make a decision. But suppose some man came to you with an airplane with its engine upside down and under the rear seat, what would be your reply—and more important—your authority for making a decision? On your decisions and those of other leading men of Detroit depends the future of Detroit in aviation. . . . I am going to send out a weekly series of letters, purely educational, giving the fundamentals of aviation—what makes a plane fly, etc.—for the education of Detroit's management. Will you have your secretary call or drop me a line if you would like to receive the series?"

From the first one hundred letters Stout received sixty-five replies from industrialists requesting the series. As he learned later, the recipients usually read the letters carefully, and some kept them on file for years.

In the fourth letter Stout enclosed a filing card and paper clip, explaining how to fix the clip and bend the card to make a crude paper airplane. Several days later at the Detroit Athletic Club, Stout met Roy Chapin, Hudson Motor Car Company president and a newsletter reader. "Bill, you got me into trouble," Chapin said. "I received your letter just as I was leaving for New York on the train. I put it in my pocket and didn't pull it out until I got into my compartment. I didn't have enough room there to fly the plane so I opened the door and let it go. Well, it flew halfway down the aisle, and soon everyone in the Pullman was outside my compartment, and I ended up giving a lecture on aviation."

Stout's letters were a prelude to a personal visit by him and his sales manager, Stanley Knauss, to each executive on his mailing list. Once inside the prospect's office, Stout would describe in detail the plans of the new Stout Metal Airplane Company. "We want to build a new radical type of transport plane using a cantilever wing," Stout would say. "And we want to build the plane entirely of duralumin, a new light metal of super strength."

Stout also played on local pride. "If Detroit expects to take its place in aviation, someone has to build planes," he said. "This city started out building the nation's first metal airplanes with me, and I now know more about them than anyone in the world."

Stout then got into the final "clincher," which almost never failed to produce results. "We want to build another metal plane. If you join us, it will cost you money. One thousand dollars. No more, no less. And for your one thousand dollars you will get one definite promise. You will never get your money back."

The first Detroit industrialist on whom this unusual approach was tried was Charles Bush, head of a wholesale hardware firm. Not only did Bush write out a $1,000 check on the spot, he also advised Stout whom to see next. Within a few months Stout had received checks from many important Detroit industrialists—Walter Chrysler, the Fisher brothers, Horace Dodge, William S. Knudsen, R. E. Olds, Albert Champion, Charles F. Kettering—the list read like a "Who's Who in the American Automobile Industry."

Then Stout expanded his list to include industrialists from other cities. Added to his list of stockholders were Philip Wrigley, Harvey Firestone, Marshall Field, Edward Budd, Gar Wood, and R. L. Polk. Then one day, to Stout's plant came William B. Mayo, chief engineer of Ford Motor Company. Mayo at that time was the most powerful man in the entire Ford organization next to Henry Ford himself and his son Edsel. A self-taught engineer, Mayo had originally worked for an engineering firm brought in to design the powerhouse for Ford's Highland Park factory. Ford's own plant engineer had said that Henry Ford's idea of installing extra-large steam-gas generators would never work, but Mayo said it would. Attracted to a man who would attempt the seemingly impossible, Ford told Mayo to go ahead and install the generators. When they operated perfectly, Ford arranged for Mayo to work part time for Ford Motor Company, and by 1915 Mayo was the company's chief engineer.

The Fords for years had quietly been exploring aviation possibilities but, like many others at the time, they leaned toward lighter-than-air craft. Mayo on the other hand was fully committed to heavier-than-air machines. For a long time he had been friendly and sympathetic to Stout, and had visited the Stout Engineering Laboratories during the building of the Navy Batwing. Mayo felt that Stout, more than any other American aeronautical engineer, was headed in the right direction. He had been one of the Stout Metal Airplane Company's original $1,000 subscribers. Now he had something else on his mind.

"Bill," he said. "I think I can arrange a meeting for you with Henry Ford."

Stout could hardly conceal his excitement. If Ford became a stockholder, it would be the biggest single thing that could happen to the Stout Metal Airplane Company. Financial support, prestige, publicity—all would follow immediately, once Ford agreed to participate.

"I'll do what I can, and let you know," Mayo said.

The next day the telephone rang in Stout's office in the Stout Metal Airplane Company.

"Bill, I've arranged it. You are to meet Henry Ford at 10 o'clock next Tuesday at the executive offices in Dearborn."

David A. Weiss

Henry Ford was America's most famous individual. His Model T automobiles had made his name well known throughout the world. More than anything, the name of Ford was synonymous with assembly-line production. Others had manufactured automobiles before, but no one else had had the vision and technical know-how to produce them in large quantity. The Ford Motor Company, with assets of over a billion dollars, had two hundred thousand direct employees and two hundred thousand indirect employees, and all by itself was the third largest industry in the United States.

There were no outside stockholders in Ford Motor Company. The Fords—Henry, his wife Clara, and his son, Edsel—owned the company themselves. They had built it up by themselves too, turning back the company's huge profits to finance new facilities, new plants, and new production lines. Never once did the Fords issue bonds to obtain financing, nor did they ever borrow from banks.

Ford's operations had forged new approaches in management and production. To insure the quality of his automobile (and also to increase profits), he had taken over the functions of most of his suppliers. Instead of buying timber, the Ford Company had its own lumber forests. It also had its own mines, railroads, glass plants, steel mills, and even a fleet of oceangoing freighters. The Ford factories had the job of making auto parts. These were shipped to thirty-five assembly plants strategically located throughout the United States. In these plants the automobiles were assembled.

Henry Ford was born on July 30, 1863, on a farm near Dearborn, Michigan, nine miles west of Detroit. The eldest of six children, he attended school in a one-room schoolhouse where he studied from McGuffey's Readers. Both after school and during vacations Ford worked hard on the farm, doing the usual chores. So hard was the hand labor, he began to think at a very early age that there certainly must be an easier way to get work done.

This thought led Ford into mechanics, an interest encouraged by his mother, who told Ford he was a born mechanic. Unfortunately Ford's father did not approve. He thought young Henry should stick to farming and forget about the newfangled mechanics. But when Ford was twelve, he became more interested than ever. That year he received a watch as a gift and he also saw his first horseless vehicle, a steam-powered road engine used to drive a threshing machine.

Ford thought first of trying to make a working model of the road engine, but he then got intrigued with the watch. He had a burning curiosity to know how it worked, and at night—secretly, so his father would not find out—he took the watch apart and put it back together again, using only two handmade tools. One was a screwdriver he had fashioned from a knitting needle, the other, a pair of tweezers he had made from a spring.

Soon Ford was so handy at watch repairing he was fixing watches for neighbors as far as ten miles away. Then, at sixteen, he decided to go to Detroit to pursue further his mechanics career. He walked to the city and took a job as a repairman for the Michigan Car Works, with a starting salary of $1.10 per day. Other jobs followed, at foundries and at machine shops, and soon Ford had mastered the mechanics trade.

At night he continued with his watch-cleaning and repairing in order to make extra money. This gave him an idea. Why not adopt machinery methods to produce watches in quantity? If he could bring down the costs of making watches, he could sell them for as little as a dollar each. Such a low price would generate mass sales, and mass profits would also be possible.

But Ford's father called him back to the farm. Ford's mother was dead by now, and his father offered him—as an inducement to return—a gift of forty acres of wooded land. Accepting the offer, Ford set up a sawmill on his new property, earning income by cutting and selling the timber. For a time he was happy. He married Clara Bryant, daughter of a nearby farmer, and with his own hands built a house on the land.

In his new home Ford set up a workshop where he started tinkering again. Watches were no longer his chief interest. He had seen in an English magazine a description of the early internal-combustion engine invented by Nikolaus Otto, and, later, he had an opportunity to repair one. Convinced that the gasoline engine, instead of the steam engine, was the engine of the future, Ford decided to build one of his own design.

Since his home workshop in Dearborn was not well enough equipped for this task, Ford in 1891 moved back to Detroit. To support himself and his wife, he took another machinist's job, but at night, in a small brick shed back of the house, he worked on his engine. Two years later the engine was completed. It was primitive. Its cylinder was reamed out of a gas pipe; its piston was homemade; and its flywheel was a wheel he had taken from a lathe. Testing took place in Ford's kitchen. With the engine clamped to the sink and its sparkplug attached to an overhead light socket, Ford started it up, as his wife looked on.

The engine worked, and Ford immediately decided to build a four-wheeled horseless carriage to install it in. He had seen such carriages in Detroit; they had been handmade, and most were imported from Europe. By 1896, Ford's carriage was complete. Its body was a light frame mounted on four bicycle wheels. Its seat was a bicycle seat. To get the carriage out of the workshop, Ford had to break down one wall. The carriage ran fine, and soon after, Ford added seats for his family. In the improved vehicle he drove back to his farm one Sunday with his wife and three-year-old son Edsel.

By now Ford had a good job. As chief engineer of the Detroit Edison Company he was earning $1,800 per year. But automobiles were his major interest. As soon as he sold his first car for two hundred dollars he began building a second. Several prominent Detroit businessmen who had watched his progress offered him $100,000 to organize a company to build ten of his cars. Resigning his job, Ford started building the first car, but he soon parted company with the businessmen. His interest was to build a really good car; theirs was to make large profits. Other businessmen then came to Ford. They too wanted him to organize an auto manufacturing company. But this company—Henry Ford Company—also broke up when Ford disagreed with his backers over policy.

Now Ford decided to build automobiles on his own. He reasoned that the way to get attention was to build fast cars, so he began the construction of two racing cars. One was the "Arrow," the other, the "999," and both had 80 hp engines that enabled them to go 60 mph. Barney Oldfield, the famous auto racer, was hired to race the autos, and Oldfield won one race after another. The "999" never lost a race!

By this time there were more and more automobiles on American roads. There was also an American automobile manufacturing industry. Companies like Oldsmobile were producing several thousand cars a year, and total United States production had reached almost ten thousand cars. With his name now known because of the racing cars, Ford decided to try manufacturing one more time. Organizing a new company, the Ford Motor Company, he secured investors. The company's total capitalization was $100,000, and Ford set up the company's incorporation in such a way that, although he had put in no money, he controlled the stock.

From the company's first automobile, its product got a reputation for being tough, well made, and simple to operate. Various models were introduced—the Fordmobile, the Model B touring car, and then, in 1908, the famous Model T. As Ford himself said, "This car contained all I was able to put into a motorcar."

The Model T was noisy and unattractive, but it was also an amazingly efficient automobile. The year of its introduction it made headlines by winning a cross-country race from New York to Seattle, covering the distance in twenty-two days. By 1911, the Model T was so much in demand that Ford had to build a new plant at Highland Park to produce it, and he had to expand the labor force to four thousand employees.

By 1913, Ford had set up the first of his famous mass-production assembly lines, and still he had trouble keeping up with orders. So profitable were operations now, he was able to buy out his original investors at a cost of millions, and in the reorganized company he made his son Edsel the president. By 1923, the year Ford met Stout for the first time, the Ford Motor Company was producing Model Ts at the rate of two million per year.

Through the years Henry Ford had been approached many times to participate in aviation manufacture, but he had turned down all offers.

Edsel, who was more interested in aeronautics, had several times prodded his father to take action in this field, but without any significant results. Actually, in 1909, only six years after the Wright brothers had flown at Kitty Hawk, Edsel had become involved in building an airplane. A childhood friend named Charles Van Auken had started to build a flying machine with several other Detroit youths. Coming to Edsel, who was only fifteen at the time, Van Auken asked for help. Edsel became interested in the project and went in turn to his father for assistance. Pleased that Edsel was involved in something mechanical, the elder Ford rented the group a barn in which to build the plane. He also had some of its machine work turned out in a Ford plant.

*The Van Auken-Edsel Ford flying machine built in 1909, **only six** years after the Wright brothers' first powered flight at Kitty Hawk.*
FORD ARCHIVES PHOTO

When completed, the boys' plane was a single-place, high-wing monoplane. Constructed of fabric, metal, and glue, it was held together by struts and wire. The plane was powered by a 28 hp direct-drive Model T engine, and its controls were of a conventional type, consisting of rudder, ailerons, and elevator. Several of the plane's features, however, were unusual. At a time when all other planes had a tail skid, this plane had a tricycle landing gear. Also, its operator did not have to fly lying prone on the fuselage, as he did in the Wright machine. In this plane he could sit upright.

Although Edsel himself did not fly in the plane, Van Auken and the others did. Their first flights were made from a field that Ford owned on the present site of the Dearborn Country Club. The plane flew, but it was obviously underpowered. Henry Ford again came to the rescue—with a new souped-up Model T engine. Using this engine, Van Auken made several more successful flights. But when the plane was taken later to the Ford parade grounds for a demonstration, it barely got off the ground, and, shortly after, it crashed into a tree and was smashed beyond repair.

Henry and Edsel Ford probably would have gone on supporting Van Auken and his plane. But about that time the Wrights made headlines when one of their new planes scored a spectacular success at army air trials in Fort Myer, Virginia. Prospects for the Van Auken plane seemed dim in comparison.

Thus the Fords eased out of this aviation project, but others came up from time to time in succeeding years. During World War I, when the need for warplanes was so desperate, the Fords proposed to the United States government a gigantic project whereby they would produce 150,000 airplanes a year on a moving assembly line. They even proceeded so far as to ask the managing director of the Ford Motor Company of England to supply them with a British or German plane to use as a production model. But nothing came of this proposal, nor of another scheme which Ford explored with Captain Henry H. "Hap" Arnold of the United States Army Air Force, to develop pilotless bombers.

Like most automotive manufacturers, Ford participated in the successful war program to manufacture Liberty engines in quantity, and in 1919, when the Ford Motor Company revised its corporate charter, a provision was inserted for aircraft production. Nonetheless, the Fords still leaned to dirigibles as the most practical form of commercial air travel. After World War I, Ford had met Dr. Hugo Eckener, the German airship builder, and had become intrigued with the possibilities of using dirigibles as commercial transports. Ford saw promise in the fact that their range and carrying capacity far exceeded those of the airplane of the day, and he had explored with Zeppelin company officials in Germany the possibility of Ford's building Zeppelins in the United States. The Zeppelin company had offered their patents to Ford for this production, but he did not want to go ahead unless he had the support of the navy.

Mayo had talked many times to navy officials, but nothing had developed. Meanwhile, Ford began looking closer to home. The head of the Goodyear Rubber Company's balloon division was an engineer named Ralph Upson, who believed the future of dirigibles lay in metal skin rather than rubber. With Goodyear committed to rubber, Upson decided to look elsewhere. Hearing of his availability, Ford helped organize around Upson a Detroit company called Aircraft Development Corporation which had as its goal the building of an all-metal dirigible. Coincidentally, the metal to be used in the new dirigible was duralumin, the same material Stout had used in his navy Batwing plane.

Ford's support of Aircraft Development Corporation made him more receptive to the Stout Metal Airplane Company, for here were two companies endeavoring to conquer the air, and both were experimenting with metal, a material that lent itself to the mass production he loved so well. There were other reasons that Ford was interested in Stout and his metal planes. In 1920, when Ford first began thinking about the possibilities in dirigible transport, he had sent Mayo to Europe to report on their use and manufacture. When Mayo returned, it was not Zeppelins he emphasized, but airplanes. As Mayo saw it, the beginnings of commercial aviation in Europe were already demonstrating the practicability of the airplane as a passenger and cargo carrier.

When Mayo later began telling the Fords about Stout, they became increasingly interested in him and his work. Ford saw in the eccentric inventor something of a kindred spirit. Like Ford, Stout was an ex-country boy working in a sophisticated industry. Like Ford, Stout was handy with tools, and he had once built his own home and furniture just as Ford had done. Like Ford, Stout had ideas that people called zany, but that somehow worked.

Many years before, Ford had written down his four criteria for success in business. One criterion was absence of fear for the future; another was the elimination of veneration for the past; another was putting service ahead of profit; the last was not worrying about competition. From what Ford had heard about Stout, the airplane inventor met all these criteria as well as Ford himself did. Ford liked to hear that Stout worked long hours. (If Ford hated anything, it was idleness.) He liked Stout's slogan, "Simplicate and add more lightness." He thought the concept of all-metal planes was excellent.

As the time of the meeting between Ford and Stout approached, Stout grew more excited. Driving to the Ford Motor Company headquarters in Dearborn on the appointed day, all he could think about was that if Ford would become an investor, it would brighten the future of not only Stout's own company, but also of commercial aviation as a whole. It was not the unlimited financial resources of the Fords that Stout was thinking of. Rather, it was the fact that Henry Ford and Edsel Ford—owners of the world's largest automobile company—would, by backing him, be also endorsing aviation at a time when it had no prominent support. In the public's eye, the entrance of the Fords into aviation would automatically give aviation something it had never had before—*substance.* No longer would pilots be thought of as daredevils and crash victims. No longer would airplanes be shrugged off as flimsy, dangerous toys. No longer would air transport companies be considered small risky enterprises. If Henry Ford participated even to the slightest degree in aviation, the fly-by-night reputation of the industry would be transformed. Henry Ford represented big business. His name was synonymous with reliability and dependability. Anyone who had ever ridden in or driven a Model T could testify to that.

Stout realized from Mayo that the Fords were interested in his company or they would never have asked to meet him. But he also knew that Henry Ford at times was unpredictable. This man, who operated one of the world's largest industrial empires, conducted his business operations in a very strange way. Unlike most American businessmen, he distrusted bankers and lawyers and avoided them as much as possible. He also kept no office hours, dictated no letters, and wrote relatively few. His staff meetings were held at lunchtime in a little white house on the company grounds where he, Edsel, and Mayo ate lunch every day they were in Dearborn.

When Stout arrived for the meeting, everyone was waiting in Henry Ford's office. Seated behind a desk was Henry Ford himself. Stout recognized immediately the wiry spare frame, lean face, and silver gray wavy hair. Sitting beside Ford was Edsel, slender, youthful, and sensitive, with his dark wavy hair carefully combed back. Also present was Mayo.

Stout began talking excitedly, but relaxed when he saw Henry Ford's sympathetic face. The senior Ford even smiled several times as Stout explained at length his aeronautical theories and his plans to build metal planes. The Fords and Mayo listened attentively, and then Henry Ford finally interrupted.

"How much money do you want?"

"One thousand dollars. No more. No less."

"We want to invest more."

"That's the limit for everyone."

Henry Ford frowned. Then his eyes suddenly twinkled, and he looked at Edsel.

"That's easily solved. I'll put in $1,000 for myself, and $1,000 for Edsel."

Stout saw no objection, and shook the hands of the two Fords and Mayo. Driving back, he kept thinking over and over. His contributions to aviation were already significant. He had many more ideas with which he thought he could revolutionize aviation. But, when all was said and done, it might be that what he had done that day—bringing the Fords into aviation—would be what he would be remembered for most of all.

For a time the Fords' participation in the Stout Metal Airplane Company was kept quiet. Then the company's board of directors was reorganized to make room for Edsel, and the word got out. Newspapermen traveled out to the Ford plant to interview Henry Ford. About Stout, Ford said only that they were backing him because they believed in his ideas and thought he could make good planes. About aviation in general, Ford was more talkative.

"Aviation will come of age when planes are commercially developed," he said. "And I can foresee the days when they will be built in the thousands . . . Airplanes will be more effective than autos in wiping out misunderstandings between people, because they will reduce distances in the world. They can even work for peace, because when they become popular they will put power in the people's hands . . . and make short work of the warmongers."

Refinanced now, Stout moved ahead with the building of the Air Sedan. Although his goal was to construct an all-metal plane, he decided to build first a plywood-and-fabric model to test the plane's aeronautical performance, and then go on to the all-metal version. This double-step approach all but exhausted the new funds he had received from backers, but he did complete both planes, each a four-place monoplane with a cantilever batwing.

From the time of the first test flight in February, 1923, it was obvious that the plane needed more power. Stout had tried to purchase a 150 hp Hispano-Suiza engine but these motors were scarce and expensive in the United States. He had to settle for a Curtiss OX-5 90 hp motor. One day when the all-metal Air Sedan was being tested at Selfridge Field, Henry and Edsel Ford drove over. Those watching the tests were not particularly impressed over the results, but Henry Ford turned to Edsel.

"You know, I am more convinced than ever that this fellow is on the right track."

Stout, standing nearby, overheard Ford's remark. "Mr. Ford, this thing won't work a damn until I get more horsepower, and to accomplish this, I need more money."

Henry Ford looked at the plane sitting on the field. "Son, you don't need more money, you need more airplane."

The ever-resourceful Stout had an answer to this. From the experience he had gained in constructing the two Air Sedans, he had already conceived a much larger plane. That afternoon at Selfridge Field, he described this new projected plane to the Fords. It would be called the Air Transport, he said, and it would not only be more graceful than the Air Sedan, but also more powerful. With all-metal construction and high-wing design, it would be built around a 420 hp Liberty engine and be able to carry a payload of a ton either in passengers (eight plus a crew of two) or freight.

The Fords were impressed. No plane of this size and power had ever been built in the United States. They decided to invest more money in the Stout Metal Airplane Company. Other shareholders also contributed more funds, and Stout started in on his biggest airplane project yet. Since it was Stout's second commercial plane (the first was the Air Sedan), it was designated the 2-AT. Stout invited each of his backers to visit the plant and drive a rivet into the new 2-AT plane. "This will be the first commercial airplane ever built in the United States, and this rivet will mean something to you and your grandchildren," he said.

Stout in mock-up of his 2-AT "Air Transport." The first all-metal commercial transport plane ever built in the United States, it was given the name Maiden Detroit *when it was completed,* FORD ARCHIVES PHOTO

As work continued on the Air Transport, Stout picked up more support from some of his fellow aeronautical engineers as far as the superiority of metal construction was concerned. However, those who resisted metal were still clearly in the majority, and at every professional meeting, Stout ended up having to defend metal. At one such meeting in Washington, D.C., Stout locked horns with Anthony Fokker. The Dutch-born aircraft manufacturer, who had been one of the leading airplane builders in Germany, had recently set up an American subsidiary that manufactured planes in a factory at Teterboro, New Jersey. Fokker's planes were built of wood and veneer, and at the meeting he criticized Stout's views, proclaiming loudly the advantages of wood construction: its cheapness, its strength, the ease with which it could be shaped and repaired.

Stout argued the case for metal. He agreed a veneer plane was as light as a metal plane. He conceded metal was more difficult to repair, and that it could be shaped no more accurately than wood. But then he told of being with Fokker one day in Detroit. A young man had come up who had just purchased a new Fokker. "It's in the crate now, but I expect to get it assembled this afternoon. Would you like to fly in her then?" the young man asked. Stout noted that Fokker had politely declined. There was a sound reason for this, Stout told the meeting. Fokker did not know the condition of the plane's wood. "You see," Stout said, "any plane built of wood starts getting a disease after six months. That disease is veneer-eal disease."

Everyone laughed, even Fokker. But the only aeronautical designers at the meeting who supported Stout's view were Igor Sikorsky and Giuseppe Bellanca.

In building his Air Transport, Stout made many technical innovations. Unfortunately, he was so far ahead of other manufacturers, he could not obtain much information from technical papers or recently published research. In Washington, D.C., the National Advisory Committee for Aeronautics (NACA) published what technical data it could. But, as Stout once remarked, the sum total of this data could be contained in a pamphlet 1/8" thick. Still, he did get some assistance from NACA. When he set out to design the wing for the Air Transport, he used an experimental contour the committee had recently developed. Like the wings of his Batwing planes, the wings of the Air Transport were also big and thick. The Air Transport's structure was similar to but larger than the structure Stout had used on the Air Sedan. For the fuselage shape, Stout settled on a configuration squared on the bottom and half round on top. For the skin, he again used corrugated metal.

Stout decided upon the revolutionary concept that the Air Transport should have a copilot as well as a pilot. This in itself provoked controversy (all other planes had but one pilot), but when it also became known that he intended to place the pilots just forward of the wing's leading edge, criticism mounted. Pilots should be back in the tail, everyone said. In case of a crack-up, they would have less chance of getting hurt.

Stout had an ulterior motive. He was designing his new plane to carry passengers, and in his opinion, the passengers were the ones who should have the most protection. Stout also felt that if the pilots were up front, better visibility could be achieved. So he placed the pilots' cockpit in front of the passenger cabin, and he enclosed the cockpit. To provide good visibility, he installed windows that could be opened, and in front, a windshield. Since shatterproof glass had yet to be developed, the material used for the windows and windshield was celluloid, the only transparent plastic then available. Bracing it were duralumin strips.

One of Stout's problems with the new Air Transport was that he needed a good motor the Stout Company could afford. This was solved when one of the $1,000 investors, on visiting the Marmon auto plant in Indianapolis, spotted in a glass display case a highly polished, nickel-plated Liberty engine. A sign next to it said that Marmon, during World War I, had won a War Department citation for Liberty engine production. When the investor found out that the engine was still operable, he asked if Marmon would sell it. "Sure, the war's long over," a Marmon executive said. "We'd junk it for two hundred dollars." Stout was overjoyed when he heard the news. The investor contacted Marmon again. "You've got a deal."

The Liberty engine needed only some re-welding and a new cylinder head. Otherwise, it was in excellent shape. Installing it in the new Air Transport, Stout soon had the plane ready for testing.

A few months before, a powerboat had obtained publicity racing on the Detroit River. It too had a Liberty engine, and it had been built by Gar Wood, a well-known boat manufacturer who was also one of Stout's investors. Gar Wood had called the boat "Miss Detroit," and this gave Stout an idea.

"Let's call our plane the *Maiden Detroit,*" Stout said. "The 'Maiden' will complement Gar Wood's 'Miss,' and it will also be a pun on the fact that it was 'made in' Detroit."

The 2-AT Maiden Detroit *on its first flight from Selfridge Field,* FORD ARCHIVES PHOTO

On April 23, 1924, the *Maiden Detroit,* first of the new 2-AT series, was towed out to Selfridge Field for its first tests. It was a windy, cool day, and ice from the previous winter still covered Lake Erie. The shiny, new, all-metal *Maiden Detroit* was pulled between the lines of old wood biplanes, and Walter Lees, the test pilot, climbed into the plane and started the engine. Taking off against the wind, he headed across the lake—when suddenly something happened. In less than a minute the plane was down, sitting snugly on the ice.

Stout and the others ran down the airfield onto the ice-covered lake. When they reached the *Maiden Detroit,* Lees had already stepped out. He pointed to the celluloid windshield. Blown in by the wind, it had been jammed against his control wheel. The only way Lees had been able to free the controls to land was to push the celluloid back in place with his foot.

Stout fixed the windshield, and in a few days Lees took the *Maiden Detroit* up again. Again the test ended in failure. A faulty carburetor forced the plane down into a nearby swamp. To get the *Maiden Detroit* back on dry land, Stout had to go out and completely disassemble the plane.

But on the third try everything worked perfectly. With sandbags in the seats simulating the weight of passengers, the *Maiden Detroit* took off carrying its full weight, and for over an hour it performed with exceptional control. So pleased was Stout that, the minute Lees landed, he and Mrs. Stout themselves went up for a flight over the city of Detroit.

In the next few weeks, the *Maiden Detroit* was taken up again and again, and Stout's big new corrugated-metal plane became the talk of Detroit. Commented the *Detroit Free Press* in an editorial hailing Stout's achievement: "[This is] Detroit's first bid for commercial aviation."

For Stout, the *Maiden Detroit* climaxed five years of spectacular innovations in airplane design. The advanced Batwing planes that Stout had built were achievements in themselves. But even more significant were his two major contributions to aeronautical design: his removal of the "parasitic resistance" of the engine, controls, fuel tanks, etc., and his use of metal in aircraft. Because of these contributions, Stout had helped make commercial aviation possible. The army De Havilland DH-4 planes with Liberty engines could carry only two people. His *Maiden Detroit*—with the same engine—could carry eight. In terms of freight, the capacity of the *Maiden Detroit* was one thousand pounds, not including the weight of either the pilots or the fuel.

Throughout the summer of 1924, Stout took up more than a thousand passengers in a series of *Maiden Detroit* demonstration flights. There was more to these flights than sightseeing. Detroit had talked about building a municipal airport, but no site had been selected. A proposed site was Water Works Park in downtown Detroit, but many tall buildings surrounded it. Stout felt an airport located in the park would be hazardous for planes. To prove his point, he took up leading Detroit citizens and showed them from the air the inadequacies of an airport at Water Works Park. He also pointed out more suitable sites. As part of his campaign, Stout invited Detroit's City Council to see the park from the air. The only member who accepted (the others were afraid to fly) was the oldest man in the council.

While everyone in Detroit procrastinated about the airport, Stout grew impatient. For each flight, he had to tow the *Maiden Detroit* twenty-five miles to Selfridge Field. Irked by this inconvenience, Stout wrote a letter to Henry Ford. He told Ford of his need for an experimental airport from which he and others could more conveniently make test flights.

Ford's answer came quickly. Stout was invited to Dearborn to confer about a possible landing field there. The Fords were very impressed with what they had been hearing about the *Maiden Detroit,* and they realized Stout needed all the support he could get. Also, still very much interested in dirigibles, they had begun thinking in terms of a combination airfield that could service both heavier-than-air and lighter-than-air craft. When Dr. Eckener, the German dirigible expert, had visited Dearborn, Ford had asked, "Why don't you fly a Zeppelin here?" "I would if I had a place to tie up," Eckener answered. "Well, we'll build you one," Ford had said.

Since then, Ralph Upson's Aircraft Development Corporation, which the Fords had also been supporting, had begun construction of a small experimental duralumin dirigible at nearby Grosse Point. It was only to be ninety feet long, but because of its metal construction, it would be faster and have a greater carrying capacity than a fabric-covered dirigible twice its size. Like Stout, Upson had developed new techniques in applying duralumin. The metal sheets for the dirigible were first riveted together, and then, after being put into place, were fastened by an ingenious stapling device. So tight were the dirigible's joints, no gas could leak out.

Carl B. Fritsche, an executive in Upson's Aircraft Development Corporation, was also asked to come with Stout to the Ford Motor Company that day to talk to Henry Ford about the landing field. Ford met the two men and escorted them around the twelve thousand acres he owned, showing them what was available. It had already been agreed between Stout and Fritsche that Stout could do the choosing.

"Well, Stout, we'll give you your airfield," Henry Ford said. "Just tell me which site you think is best."

Of all the tracts Stout had seen, only one had enough space for the thousand-foot runways that Stout figured were necessary.

"The land near the laboratories and Oakwood Boulevard is best, I think," Stout said.

A Ford secretary walking with the group spoke up. "But, Mr. Ford, we had set aside that land to use as a subdivision for homes we were going to build for workingmen."

"That was yesterday," Henry Ford said to the secretary. "Today it is a flying field."

Ford then turned to Stout. "We want to make this field the best landing field in the world."

Stout was overjoyed. He knew Ford meant what he said. Then Ford had another surprise for Stout and Fritsche. The field Stout had selected was over six hundred acres in size. Little more than two hundred acres would be needed for the airfield itself. On the rest of the land Ford offered to construct a factory that could be used by both the Stout Metal Airplane Company and the Aircraft Development Corporation.

Stout and Fritsche could not have been happier. Both companies had been operating in small, limited space. The new building that Ford was proposing would solve many of their production problems.

A few days later the press got hold of the story. Interviewed, Henry and Edsel Ford explained they were donating the land for the field and building the factory as "an incentive to the development of metal aircraft construction in Detroit." They commented favorably on the all-metal plane *Maiden Detroit* just completed by the Stout Metal Airplane Company as well as the "metal super dirigible" being worked on by the Aircraft Development. Corporation. Edsel emphasized they merely wanted to help these two companies develop, and that the Fords "had no intention of entering the airplane industry."

Even so, the Ford move was hailed by the press as a major development. A modern airport at the Ford Motor Company headquarters in Dearborn! A factory where not only metal airplanes would be built, but also metal dirigibles! Aviation had not had such a boost since World War I, when the auto manufacturers had gotten together and mass-produced the Liberty engine.

The Fords suddenly were in the aviation limelight. Continually the press asked Henry and Edsel to comment on various aeronautical developments in the United States and abroad, and usually the Fords had ready answers. They even bylined articles on aviation in leading magazines. In an article Edsel wrote at this time for *Nation's Business,* he predicted that "Aircraft will soon play an important part in the commercial life of the United States. I believe the dirigible will handle transatlantic service, and the airplane, service within the United States. There is no doubt in my mind we have entered the age of aviation."

Chapter Four
Made in Dearborn

Henry Ford wasted no time in building the airfield he had promised Stout and Fritsche. Within twenty-four hours after Stout had selected the airfield site, hundreds of men, scores of trucks, and almost forty Fordson tractors were at work clearing, leveling, and grading. Ford himself paid a visit almost every day, and spent hours there talking to engineers and workmen.

One day he happened to talk to an airplane pilot who had driven up from Detroit to see the progress of the new airfield. "These high tension wires," the pilot said, pointing to the electrical lines behind the field, "they'll be dangerous to anyone trying to land." That was all Ford needed to hear. The next day he ordered his plant engineers to bury the lines in underground conduits. The project cost thousands of dollars.

On another occasion, when the field was partially usable, some army pilots flew in from Selfridge Field. Not far from one end of the field was a grove of more than one hundred maple and walnut trees. Ford loved the trees, but he could see that some planes had difficulty clearing them. Again he did not hesitate. He ordered them chopped down. "I don't want to see any of these young fellows killed," he said.

But when they started laying the factory's foundations, it was discovered the building line touched a crab apple tree. "There is no use destroying something that takes so long to grow," Henry Ford said. This time he ordered the factory building moved three feet away.

Soon after, Stout was asked what kind of factory building he wanted. Appreciative of all that the Fords were doing, and not wanting to seem overly demanding, Stout selected the most inexpensive type of steel and brick construction. But Ford hired one of America's highest-priced

Opposite: Henry Ford and William B. Stout in one of the rare photographs taken during the early days of the Stout Metal Airplane Company. FORD ARCHIVES PHOTO

engineering firms to design the building. When Stout said such expense was not necessary, Ford told him not to worry. He said that when Stout's rent was determined later, he was sure Stout would find it reasonable. Shortly after, someone from the Ford organization telephoned Stout and asked if a rental of $150 per month for the factory would be too much. The figure was so low that Stout thought it was a joke, but he did answer, "No, it would not be." Stout never found out who had called, but that was all the rent he ever paid.

Meanwhile, from the Detroit factory of the Stout Metal Airplane Company, Stout continued his missionary work with the *Maiden Detroit.* Besides taking up prominent Detroiters on short sightseeing flights, the plane took emergency trips, many of which ended up publicizing the value of the plane as a transport. Near Oscoda, Michigan, the army was holding exercises when a shortage of ammunition suddenly developed. The *Maiden Detroit* saved the day by bringing in new ammunition supplies from Selfridge Field.

Then, in October, 1924, at the National Air Races in Dayton, Ohio, the *Maiden Detroit* was used to shuttle important Detroit industrialists back and forth to Dayton. Eddie Stinson was the pilot, and Stout and George Prudden usually went along. On one flight, the plane, counting Stinson, carried a record ten passengers. One was Professor Edward P. Warner, of the Massachusetts Institute of Technology, recently appointed by the United States Post Office to investigate new types of aircraft for the Air Mail Service. About two-thirds of the way home, the engine of the *Maiden Detroit* suddenly caught fire. Flames shot over the celluloid windshield. Stinson, about fifteen hundred feet up, quickly put the plane into a vertical sideslip dropping one thousand feet suddenly to blow out the flames. Then he reversed the plane's direction another one hundred feet—just to make sure the fire was extinguished.

In all this maneuvering, Stinson had lost wind direction and the engine quit. There was nothing to do now but land, but when Stinson looked down, there was no field to land in. Rather, there were two fields, separated by a fence. Stinson saw his only way out was to hedgehop.

Unfortunately, Stinson undershot the first field about six inches, and a fence post caught on the plane's tail, tearing it off. With no controls left, Stinson crashed through the fence at 70 mph, scattering broken fence rails in all directions. In the process the *Maiden Detroit's* landing gear was ripped off, but the plane, sliding in on its belly, made it safely into the next field, where it stopped right side up.

Everyone suffered a few minor bruises and scratches, but nothing serious. Stout, remembering how the crash of the navy Batwing put him out of business for over a year, feared the worst. This time, however, everything worked out well. Professor Warner, the new Post Office appointee, was most impressed. "I am convinced that if this plane had not been constructed of metal, we all would have been killed," he told the Post Office. Colonel Paul Henderson, in charge of the Air Mail Service, had always thought highly of Stout's planes. Upon Professor Warner's recommendation, a decision was made to purchase the *Maiden Detroit* for the Post Office. Stout arranged to bring the plane back to Detroit where the necessary repairs were made, and in December, 1924, the Stout Metal Airplane Company delivered its first plane to a customer.

Shortly after, on January 15, 1925, Ford Airport—as it was to be called—was officially dedicated. On hand were the two Fords, Mayo, Stout, Fritsche, and Upson. The highlight of the day was an aerial circus performed by a squadron of army planes from nearby Selfridge Field. Commanded by Major Carl A. "Toohey" Spaatz, the planes flew in and began looping, barrel rolling, and hedgehopping. Afterward, everyone adjourned for refreshments to the little white house where the Fords and their executives lunched every day.

An aerial view of Ford Airport when it was completed in 1929. At the time, it was the largest and most modern airport in the United States. FORD ARCHIVES PHOTO

Although Ford Airport began to be used shortly afterward, it was not fully completed for several more years. In all, the Fords would spend an estimated four million dollars on their airport, which would have not only the first concrete-paved runways in the world but also a passenger terminal building (another "first" for the United States), a restaurant, hotel accommodations, radio shack, weather bureau, traffic-control center, hangars for visiting planes, shops where spare parts and service could be obtained, and a dirigible mooring mast (the only privately owned mast in the nation). By that time the airport would be one of the world's finest, and superior to any in the United States.

Even on the day of its dedication, Ford Airport outstripped all other American airports. The two runways, unpaved as yet, were twenty-eight hundred and twenty-six hundred feet long, positioned to cross each other to take maximum advantage of wind direction. So that the field could be easily located from the air, the name of FORD was spelled out in huge letters of crushed white stone on a background of green grass.

Henry Ford's own idea, these letters were visible from a height of ten thousand feet. For night flying, the new airport had floodlights marking off the field's boundaries and grilled lights in the runways.

Ford Airport also had its dirigible mooring mast. Built on top of a tower more than two hundred feet high, it had elaborate accommodations, and an elevator to bring down passengers.

On dedication day, Henry Ford issued a standing invitation. All flyers—army, navy, marine, airmail, commercial, and private—were welcome to use the field. Ford also invited any and all dirigibles to hook up at the new mooring mast, and he announced that one dirigible had already accepted an invitation. The navy's great *Shenandoah* airship would visit Dearborn on her forthcoming mid-western trip.

Soon after the airport opened, the factory was ready for the Stout Metal Airplane Company and the Aircraft Development Corporation. It was a large one-story building, two hundred feet in length, with steel walls and a wooden roof. Inside were not only plant area, but also offices for executives and engineers. Stout had already decided his next move should be to build another 2-AT similar to the *Maiden Detroit* sold to the Post Office. The new factory would be just the place to start construction, and, giving up his Detroit plant, he moved his jigs and machinery into the new building. He already had a name for the new plane—*Maiden Dearborn.*

Both Fords visited the new headquarters of the Stout Metal Airplane Company frequently, and they told Stout they were pleased with his progress. One unusually warm February day in 1925, Henry Ford stopped by, and, somehow, instead of staying inside the factory, he and Stout ended up sitting on a pile of lumber outside. Stout was glad to have the opportunity to talk to Ford. Always one jump ahead, he had been thinking of the economics of aviation, and he wanted to tell Ford his views.

"Mr. Ford," he began, "it has always been my belief that the financial gain in aviation will come not from the manufacture and sale of aircraft, but from transporting merchandise and passengers in scheduled airline operations."

Ford listened intently. He well knew that, considering all the money put into the Stout Metal Airplane Company, the price Stout intended to charge for succeeding "Maiden" or 2-AT planes—$20,000 to $25,000 each—would never result in much profit. Even if the planes were mass-produced, the profit picture was poor. Few, if any, American commercial lines could afford to purchase them, or even the planes manufactured by others, which cost half as much. The airlines at that time had neither the capital to finance the purchase of airplanes nor the operations to justify such purchasing.

"The reason the state of the industry is so low is that it has not had a plane with a payload like ours that would make commercial operations profitable," Stout said.

Ford continued to listen.

"It we could operate a network of airlines using our planes, we could prove that an airplane can support itself economically as well as technically."

Henry Ford rose. "Stout, I think you are right, and it looks to me that what you are talking about is something somebody has to put a lot of money behind if aviation is to become an industry." He paused. "I don't know why the Ford Motor Company shouldn't do just that. Go talk to Mayo, and see what we can do about buying your company outright."

Realizing this was an opportunity to make all his dreams about an aviation industry came true, Stout lost no time in contacting Mayo. After the chief engineer talked to both Fords, he came back to Stout. Everything could easily be arranged for the Fords to acquire the Stout Metal Airplane Company. There was just one stipulation. The Fords wanted complete ownership, or none at all.

Stout immediately saw problems. What if some of his stockholders refused to sell? Why wouldn't the Fords be satisfied with a controlling interest?

Mayo shook his head. "Mr. Ford has had nothing but bad experiences with minority stockholders."

Stout, having lived in Detroit many years, knew about most of these bad experiences. Ford had had trouble buying out some of the original investors in the Ford Motor Company. Later, when he saved the Leland family from financial disaster by purchasing their Lincoln Motor Car Company, the Lelands sued him. More recently, he had been harassed by the minority stockholders of the Detroit, Toledo, and Ironton Railroad, which he had recently acquired.

Stout said he would do his best, and for several days he thought of how to broach the subject to his stockholders. Then he returned to Mayo with a proposal. The Ford Motor Company, Stout said, could take over the manufacture of the aircraft, but he (Stout) should be allowed to form an independent company that would operate an airline. All stockholders in the Stout Metal Airplane Company, including the Fords, would be invited to become stockholders in the new airline company.

"If I can offer my stockholders an arrangement like this, I think I can get them to sell out," Stout said.

Mayo liked the idea and so did the Fords. It was agreed that Ford would pay $2,000 for every $1,000 that had been invested.

"I have to break my promise to you," Stout wrote to his list of stockholders. "I said originally you would not see your money again, but now I have to tell you it will be doubled. But it will depend on everyone's selling out."

In his letter, Stout explained why the new arrangement made sense. To continue manufacturing 2-AT planes would be impossible without each investor putting up many times his original investment. However, with Ford taking over, the financial risk would be eliminated. For the first time in American aviation there would be big money and big business involved, Stout said.

"So, let me get for you $2,000 for every $1,000 you invested, and let me use the money to start a passenger airline out of Ford Airport," Stout went on. "It will be the first of its kind in the world, and without increasing your original investment, we will be giving Detroit another boost."

As Stout expected, most stockholders were only too happy to allow Ford to buy their stock and then to reinvest their $2,000 in the new airline company. But also, as Stout feared, two stockholders held out.

One, a professional minority stockholder, finally turned his stock over ungraciously in face of pressure put on him by Stout's friends. But the other refused to go along, telling Stout he was doing this for Stout's own good. "Bill, I'm your friend, and I'm telling you," he said, "the Fords are going to take you and everything you have and then after they wring you dry, they will throw you into the street. I'm not going to stand for it. You can't get my stock."

Stout tried everything he could, but the man would not budge. Finally, deciding to let the situation cool a while, Stout returned to building the new *Maiden Dearborn*.

The Maiden Dearborn *under construction. This was the first 2-AT manufactured in the new Stout Metal Airplane Company factory at Ford Airport.* FORD ARCHIVES PHOTO

As it neared completion, Mayo approached Stout and asked if he had any plans for the corrugated-metal plane.

"Yes, it's for our airline if I can ever get that one man to sell his stock," Stout said.

"Well, since that situation is not cleared up, maybe you would sell the plane to the Ford Motor Company," Mayo said.

Stout saw no reason not to sell, and the purchase was agreed on for $25,000.

Henry Ford and Edsel were pleased. They had taken up Stout's idea of an airline, but in a different way and for a different reason. They felt that something had to be done to promote public confidence in aviation, and they believed they could accomplish this by forming their own private experimental airline. This airline would transport Ford mail and automotive parts to the various Ford plants, and it would be the first in the United States to run flights on regular schedules. But transporting Ford parts was not to be the airline's primary purpose; rather, it would give prospective airline operators some idea of the costs and possible profits in commercial operations. The airline would also do much, the Fords felt, to offset the public's opinion of air travel as something dangerous.

In March, 1924, the *Maiden Dearborn* was ready for test flights. Like the *Maiden Detroit,* sold to the Air Mail Service, this plane was all-metal and carried eight passengers. However, certain changes had been made. The *Maiden Dearborn* had a narrower cockpit. It also had larger landing wheels, and it was the first plane to have the famous Ford insignia painted on its wings and fuselage.

The tests at Ford Airport were most successful, and the newspaper reporters who had witnessed them praised the corrugated-metal plane and its features. Calling attention to its all-metal construction, they pointed out that the only nonmetals used in construction were the plane's upholstery and a few minor accessories. "It is made of duralumin," they said, "a metal almost as light as paper and as strong as steel."

They also reported that the *Maiden Dearborn* was five hundred pounds lighter than any of the commercial airplanes in Europe, and one thousand pounds lighter than an average automobile. "In the air it looks like a huge silvery bird," they said, "and on the ground it looks gigantic, measuring sixty feet from wing tip to wing tip and forty-six feet long."

Even the plane's technical features were described in newspaper articles: the thirty-two-inch thickness of its wings; the design of its nose (engine, hood, and instruments, any of which could be removed and replaced within twenty minutes); and its heating system, which had the feature of warming the plane's wings to retard the formation of ice.

Henry Ford himself was as pleased as the reporters, and he told Mayo to order four more planes.

Meanwhile Ford was impatient to launch the Ford airline.

"How soon can we start?" he asked Stout one day. "I want to begin flights between Detroit and Chicago."

Stout was caught off guard. He was still testing the *Maiden Dearborn* before handing it over to Ford, and in the midst of the tests, the plane's engine had worn out.

"Can we start next Monday?" Ford asked.

Stout shook his head. "That's impossible. The plane needs a new engine."

Ford turned and started to walk away.

Sensing Ford's disappointment, Stout decided he would get a new engine installed, hell or high water.

"But we can start a week from Monday," Stout called after him.

The inaugural flight of the Ford Air Transport Service took place as scheduled on April 13, 1925. Present were Mr. and Mrs. Henry Ford, Edsel, Mayo, and Stout. The *Maiden Dearborn,* its corrugated metal all shiny and new, sat proudly on the runway. Up to the plane, wearing goggles, a helmet, and a flying suit, stepped pilot Edward G. Hamilton, recently of the Royal Air Force. Behind him was his only passenger, his mechanic, similarly outfitted.

First everyone posed for photographs. Then a few freight pieces were loaded aboard. Most of the cargo, carefully weighed, had already been placed inside the plane. It consisted of 782 pounds of auto machine parts, 107 pounds of motion picture film, and 116 pounds of other articles. Since the mechanic and his outfit weighed 208 pounds, the useful load of the plane exclusive of Hamilton and the gasoline was 1,213 pounds.

Moving into the pilot's cabin, Hamilton tried the controls. Then a Model T truck, speeding across the airfield, stopped nearby. The driver stepped out and handed small packets to Mrs. Ford, Henry Ford, Edsel, Mayo, and Stout, who passed them to Hamilton to be flown to Chicago. At 9:22 A.M., a ground crew started the engine, flipping it over by hand,

and the plane taxied down the field to get into position for the takeoff. Two minutes later, the *Maiden Dearborn* was airborne. After circling the field once, Hamilton opened her up and headed for Chicago. Henry Ford and Edsel on the ground below watched the plane disappear in the west.

Hamilton flew first to Bryan, Ohio, where he picked up the Air Mail Service's route to Chicago. Two hours and thirteen minutes from the time he took off, he reached Maywood Field, the army landing field near Chicago. A throng of businessmen and civic officials were on hand, including Judge Keneshaw Mountain Landis, the Commissioner of baseball. When they saw the *Maiden Dearborn* appear out of the eastern sky and circle the field, they broke into cheers. Hamilton brought the plane down and rolled it to a stop. A Ford truck on the field came out to pick up the freight, which was rushed to the local Ford plant on the Calumet River. Hamilton did not waste a minute. As soon as the plane was loaded up again, he took off for Dearborn, making the return trip in under two and one half hours.

The story of the Ford Air Transport Service's inaugural flight hit the front pages of the nation's newspapers. "Commercial aviation on a regular schedule began in America today with the round trip of the Stout metal airplane, *Maiden Dearborn,* which carried 1,000 pounds of freight between the Detroit and Chicago plants of the Ford Motor Company," the lead article began in *The New York Times.*

Quoted at length were Henry Ford's and Edsel's remarks upon the occasion. "This is the beginning of a new form of transportation for the company," said Edsel. "Eventually we hope to link up with our air service various Ford plants all over the midwest."

Edsel also pointed out that the service was strictly a Ford enterprise. "It will carry Ford freight only," he said.

Edsel emphasized that the Ford Air Transport Service was part of the overall Ford program to stimulate civil aviation. "We started this not only to reduce the cost of mail and express between Ford plants, but also to afford practical facilities for testing aviation equipment we may produce," he said. "We are going to tell the public what happens from stage to stage as we go along."

Henry Ford, in talking to the newspapermen, gave all the credit to Edsel. "He is the pioneering spirit behind our flying activity," he said. To this *The New York Times* added a comment of its own: "It is no secret that Henry Ford wants his son Edsel to become the same figure in aviation that he himself has become in the automotive field."

From that day on, nights were flown daily between Detroit and Chicago. But with only one plane, only one flight could be made each day from each city. Stout tried to speed up production to furnish more planes to Ford—and also to try to lower costs. Henry Ford was a frequent visitor, and helped Stout considerably with production problems. At one point Ford became so involved that he was spending half his time at the airplane factory. The other half he spent in his laboratory. Here he would draw designs on the blackboard and give illustrated chalk talks to his engineers. Then he would jump into his automobile and ride over to the airplane plant to see what was going on.

Stout was still following his formula: "Simplicate and add more lightness." By now he had also become an expert on aircraft production and had developed a methodology for building planes. First he would test his material. Then he would work out his design to fit the material. Then he would make his first drawing, usually on a scale 1/4" to a foot. After this, the jigs—the devices used to maintain the correct position between the various spars, sheets, and braces during assembly—would be welded up. Most jigs Stout built himself, and he also developed a special method for riveting metal together.

The manufacturing process started with duralumin sheets which were transferred from heat-treating baths to machines that cut out the patterns from them and pressed them into corrugated form. These pieces were then moved over to the jigs, which were permanently set in the concrete floor so there could be no variations in size. Here the sheets, spars, and braces were riveted together by means of compressed air and electric tools. So simple were the manufacturing operations, they could all be performed by non-skilled personnel.

After the wings were fabricated, they were swung over and set onto the body of the plane, the U-shaped spars of duralumin extending clear through the upper part of the plane's body. Next came the installation of the engine. As it was locked into the framework of the nose, the hood was fastened in place with just four bolts. Then the pneumatic tires were put on, and when they were inflated they raised the plane just enough to jog it off the jigs and automatically free it, as a steamship slips from the ways. The engine was then started, and the plane actually run off the factory floor under its own power.

When the second 2-AT plane built at Ford Airport—the *Maiden Dearborn II*—was nearing completion, it became obvious that there would be no room in the plant for it, once Stout started construction on the next plane. Henry Ford realized this first, and, as soon as he did, he walked down the side of the airfield a distance of several hundred yards and marked off a site for a new hangar. Here planes would be stored. Here too they would be serviced when in operation. The hangar was a permanent structure of steel and brick, and later a second one was constructed to house Ford Air Transport Service planes and various maintenance and service shops. As was the entire Ford Airport, the facilities were available to all flyers, and transient aircraft were always welcome.

The Maiden Dearborns *I and II at Ford Airport,* FORD ARCHIVES PHOTO

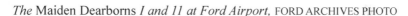

As soon as the *Maiden Dearborn II* was tested, arrangements were made to fly it on the Detroit-Chicago run of the Ford Air Transport Service. As Ford had ordered, the private airline was operating on a rigid schedule. From Chicago, a plane left every morning at 8 A.M., arriving at Ford Airport at 11:40 A.M. From Detroit, a plane left every afternoon at 3:15, arriving in Chicago at 5. Service was daily, and Henry Ford himself kept close watch on the timetable. Once when Edsel was trying to meet the plane to Chicago, he was unavoidably detained twenty minutes. Arriving at Ford Airport, he found the plane had already taken off. "This is an airline, not a yacht," Henry Ford told him.

By the end of June another 2-AT was nearing completion—the *Maiden Dearborn III*—and on July 1, 1925, this plane was added to the Ford Air Transport Service fleet. Service was now extended to another city—Cleveland. Every morning the new Cleveland plane left Detroit at 10:40 A.M., arriving in Cleveland at 12:15 A.M. In the afternoon it departed from Cleveland at 2:30, arriving back in Detroit at 4:05.

There was no question now that the private Ford airline was an unqualified success. Any visitor to Ford Airport could see that, as the three *Maiden Dearborns* shuttled auto parts and mail back and forth to the cities of Detroit, Chicago, and Cleveland. True to their promise, the Fords made public the service's operational data, and issued periodic reports on both flight performance and operational costs. The percentage of completion of scheduled flights; the safety record; the adherence to schedules; the salaries of personnel; the cost of gas, mechanical labor, oil, depreciation, insurance, and plane and motor maintenance—all this was released to the press.

Also contained in these periodic reports of the Ford Air Transport Service were statistics on the cost of flying freight and mail. As of June 2, 1925, it was calculated that each pound flown air express from Detroit to Chicago cost Ford 8 1/2 cents. As for mail, the company dispatched daily an average of 32,000 letters weighing 800 pounds. Ford statistics showed a savings in postage of $ 1,250 per day.

The same day that service was inaugurated to Cleveland, Stout made an appointment to see Mayo. He had finally gotten back all the Stout Metal

Airplane Company stock. The one man who was holding out for Stout's benefit finally gave in. In desperation, Stout had written him: "Have you enough confidence in me personally and in my knowledge of what's going on? Enclose your stock in the attached envelope and send it back to me today." Stout received the stock in the following day's mail.

It took a month for the transfer of stock to be legally effected, but meanwhile the press began playing up the story. As they saw it, aviation would come of age now that the Fords were directly involved. The *Detroit Times* wrote: "If the Fords, father and son, undertake the building of all-metal planes . . . one of our nation's greatest aviation problems will be solved." Commented the noted Hearst editorial writer, Arthur Brisbane: "The Fords have the power, knowledge, industrial genius, and money to put this nation ahead of all others in air defense. Let them do that, and their fame will outlast the memory of war."

On July 31, 1925, the day after Henry Ford's birthday, the Stout Metal Airplane Company officially became the Stout Metal Airplane Division, Ford Motor Company. Edsel was president of the new company, and Mayo and Stout were both directors. Nominal head of operations was Mayo, chief engineer of all Ford companies and divisions. But Stout was still the man in charge, and George Prudden still the chief airplane engineer.

It's official. Stout's Metal Airplane Co. Joins Ford Motor Co.

The company was shut down three days for inventory, and when operations resumed, activity was much the same as before. Meanwhile, the press came out for more interviews. Henry Ford explained again the reason he had taken over the Stout Company. "We want to do experimental work and study. We can do this better in our own plant." As Ford explained it, much had to be done in aviation despite the success with the 2-AT planes. Flying, in his opinion, was still 90 percent luck and instinct, and 10 percent science. He wanted to reverse these percentages.

Ford also announced that now he intended to expand all operations. He would accelerate the production of planes and he would extend the Ford Air Transport Service to cities as far away as Saint Louis, Indianapolis, and even New York. "I feel about aviation as I did about autos thirty years ago," he said. "It is now or never to get hold of commercial aviation and make it go."

Before he finished his remarks, Ford emphasized the role of Edsel. "The purchase of Stout's company was his idea, and he deserves the credit," he said. "There is a wonderful future in aviation, but it belongs to another generation."

The official entrance of the Ford Motor Company into aviation sparked editorials in most of the world's leading newspapers. *The New York Times* wrote: "After lagging far behind in the world race for air supremacy, the United States appears to be embarking more extensively than other nations in commercial air programs." In Europe, Ford's move into aviation was looked upon as one of the most important events since the Armistice.

Howard Mingos, a leading aviation writer of the time, authored a full-page article in *The New York Times* on the Fords and aviation. "The man who revolutionized the automobile industry has assigned himself the task of popularizing the flying machine. Their friends say that Henry Ford and his son have a vision of filling the skies with Ford planes. Henry and Edsel are undertaking the job together believing the time has come to realize the limitless possibilities of flying. They plan to use their own industrial organization, one of the world's greatest, to promote aerial transport. They say they are going to see that commercial aviation is successful. Their plan is to 'flivverize' the air."

Six months before the Stout Metal Airplane Company became a part of Ford Motor Company, the federal government enacted legislation that would prove a boon to both aircraft manufacturers and air services. This was the Kelly Act, which became law on February 2, 1925. It was designed to encourage air services to develop their facilities for passenger and freight service, and in doing so to gain the capital and experience to expand their companies. The "encouragement" was in the form of subsidies to be given for carrying United States airmail.

Until the Kelly Act, the Post Office Department had operated the Air Mail Service. Since 1918, when this service had been inaugurated, it had flown its own planes and used its own pilots. According to provisions of the Kelly Act, the Air Mail Service would continue to fly the main Columbia Route between New York and San Francisco, but feeder lines would now be flown by commercial air services selected on the basis of competitive bidding. To be eligible to bid, an airline only had to be in the air service business.

Stout and Mayo saw in the Kelly Act an opportunity to sell their 2-AT planes, which they considered ideal for passenger-freight-mail operations. With this thought in mind, they traveled to Washington, D.C., to talk to Colonel Paul Henderson of the Air Mail Service.

Henderson gave them discouraging news. Most air services expected to bid on feeder-route contracts would be hiring, as the pilots to fly their planes, the ex-Air Mail Service pilots already flying the routes. And these pilots preferred World War I-vintage DH-4s to the larger 2-ATs. The reason, Henderson said, was that, in the pilots' view, the 2-ATs were too big and cumbersome. It was true the DH-4s could carry only five hundred pounds of payload as compared to the 2-AT's one thousand pounds. But the pilots felt the DH-4s maneuvered better and gave them a greater sense of safety, particularly during night flying and under bad weather conditions.

If the Ford Motor Company wanted to sell airplanes to the air services flying the mail, it would be advised to build a new airplane, Henderson said. Such a plane should be faster than the 2-AT, should carry a bigger payload, and should be multi-engined. A three-engined plane, Henderson said, would provide extra power. It would also enable the plane to keep flying even if one engine, or possibly two, failed.

"If the Ford company would develop such a plane," Henderson said, "it would dominate aircraft manufacture in the United States."

When Mayo and Stout returned to Dearborn with this information, they found that both Henry and Edsel Ford agreed with the idea of trying to build a better plane. Henry Ford in particular saw the advantage of a three-engined plane. Not only would the extra engines give the additional power and speed needed, but they would also provide the safety he was very much interested in. "That is what we must do," he told Stout and Mayo. "We must build a three-motored plane that can fly on two motors, and, if necessary, pick itself off the ground, and even land, with only one motor."

The feasibility of developing such a plane depended on the development of a new engine. All 2-ATs used Liberty engines, and this engine with its 200 hp was still the most powerful available, even though it had been developed during World War I, almost a decade previously. But a new trimotor plane using the Liberty engine would be impractical, since, being water-cooled, it required a radiator, tubing, and other components that made its weight excessive.

Fortunately, while Stout and Mayo were brainstorming the new plane, the Wright Aeronautical Company was preparing to introduce a radial air-cooled engine. Designated the J-4 engine, this was a crude engine with push rods and rocker arms outside in the air. But with only half the weight of the Liberty engine, it not only delivered the same horsepower, it also used less gasoline and oil.

Wright offered to ship three of these new engines to Ford for experimental purposes, and once they arrived, Stout and Mayo started to design the new plane around them. Stout was the project's chief engineer. Assisting him were George Prudden and Tom Towle.

The practical Stout, in designing the new three-engined plane, followed the same concept Henry Ford had been using for years in connection with the Model T automobile, namely, making all improvements within the basic structure of the original model. Stout approached the new plane as a modification of the 2-AT—to which he would add two more engines and otherwise make as few changes as possible. He planned no radical alterations in size or structure, and hoped to be able to use most of the jigs and fixtures he had developed for the 2-AT.

Henry Ford, visiting the airplane factory frequently, became most enthusiastic about the new plane. In September, 1925, Colonel William "Billy" Mitchell, the United States Army's chief advocate for air power, came to Detroit to discuss aviation with Ford. Interviewed after the meeting, Mitchell said that Henry Ford had stopped thinking about automobiles and was devoting all his time to the company's new trimotor plane.

Ford was however not too busy to explore many other opportunities in aviation. Almost weekly, news items were reported in the press: Ford was planning a gigantic airline that would operate from Detroit as far south as San Antonio, Texas; Ford was considering establishing air service between Brazil and the United States; Ford was about to sponsor the transpolar flight being organized by Arctic explorer Vilhjalmur Stefansson—rumors flew from every direction.

Ford's involvement in lighter-than-air craft was also noted in the press. It was reported that on behalf of Aircraft Development Corporation he was negotiating for a London group to purchase an all-metal dirigible and that he had also begun discussions with the navy about contracting for a similar dirigible.

The navy was having problems with its famous *Shenandoah* dirigible, the first large rigid dirigible ever built in the United States. The huge aircraft, described as the "Queen of the Skies," had caught the public's fancy and, since its construction in 1923, had made fifty-seven successful flights. Trouble had resulted because the huge airship had become a political football. Pressure had been put on President Coolidge to approve the vessel's flight over the North Pole. In anticipation, mooring masts had been erected all over the United States and Alaska—even on ships at sea—just in case the *Shenandoah* en route would decide to descend.

But the navy felt such a flight was risky. Both its officials and the *Shenandoah's* captain, Lieutenant Commander Zachary Lansdowne, were reluctant even to take the airship on a projected nationwide tour that was to precede the North Pole flight. The first leg of this tour was scheduled for the midwest in late summer of 1925, and Lansdowne had succeeded in postponing this twice, pointing out that seasonally bad weather could be expected in the states of Pennsylvania, Ohio, and Indiana, over which the *Shenandoah* was supposed to fly.

The wreck of the Shenandoah *where it settled on a farm near Ava, Ohio.* NAVY DEPARTMENT PHOTO NO. 80-MS-11 IN THE NATIONAL ARCHIVES

The voyage was finally ordered over Lansdowne's objections. On September 2, 1925, the *Shenandoah* would start on its midwestern tour. She would leave her hangar at Lakehurst, New Jersey, fly to Scott Field in Saint Louis, jump from there to the State Fair in Minneapolis, and then go on to Detroit where she would tie up at Ford Airport's private mooring mast. While in Detroit, the *Shenandoah* would cruise over the city with prominent local citizens as passengers. Then she would return to Lakehurst, after which she would start out again on another leg of her national tour.

The Fords waited excitedly for the *Shenandoah* to arrive. They were even considering flying back on the giant flying ship to Lakehurst. An invitation to do so had been tendered by Navy Secretary Curtis D. Wilbur.

On schedule, the *Shenandoah* left Lakehurst with forty-one officers and men. But on the night of September 3, only a day and a half later, the huge dirigible, flying over Ohio, ran into severe headwinds. Suddenly, at 5:30 A.M., just north of Ava, Ohio, it hit a seventy-mile-per-hour thunder squall. Forced up to an altitude of seven thousand feet, the *Shenandoah* began to break apart into three sections. Her cabin, torn loose from its fastenings, dropped to earth like a stone. Everyone inside the cabin was killed instantly except Lieutenant Charles E. Rosendahl, the airship's second in command, who saved himself by stretching for and clinging to a girder under a gas bag as the cabin began to plummet. Others managed to escape death too when their sections, still buoyed up by the ship's remaining gas bags, slowly drifted to earth. The *Shenandoah* giant nose sailed like a balloon for twelve miles, landing safely those of the crew inside. In all, twenty-seven officers and men survived, but fourteen, including Lansdowne, were killed.

The nation was shocked, and the tragedy had direct repercussions not only for the Fords, but also for the Stout Metal Airplane Division.

Chapter Five
The Tin Goose Is Hatched

The immediate effect of the *Shenandoah* tragedy upon the Fords was to call a halt to their activities in promoting lighter-than-air craft. The United States Navy, which was considering the purchase of an all-metal dirigible from Aircraft Development Corporation, quickly broke off negotiations. Ford and Fritsche tried to explain that the safety factor in a metal balloon was three times that of the ill-fated dirigible, but the navy would not listen. It said it needed to reevaluate its entire lighter-than-air craft program. Until that was done, it was not interested in a new dirigible.

A far more serious repercussion affected the Stout Metal Airplane Division. As a result of the *Shenandoah* disaster, Stout was forced to fire George Prudden, his chief airplane engineer. Prudden, whom Ford had assigned the job of reporting on the airship disaster, had foolishly allowed the *Detroit Free Press* to photograph him writing the report at the scene of the tragedy. Henry Ford blew up when he saw Prudden's photo in the paper. The Ford company had long had a rule that no one but the Fords and a few top executives like Mayo should ever seek publicity. Prudden had not only violated this policy, he had also linked the name of Henry Ford to the *Shenandoah.*

The *Shenandoah* tragedy turned the Fords back to heavier-than-air craft. But while they waited for Stout to finish designing the new trimotor plane, they began to think of what to do to make commercial aviation more acceptable. They had long realized the importance of changing the image of commercial aviation. The flying of passengers and freight could never succeed, they felt, until the American public stopped thinking of aviation

Preceding Page: The Ford Motor Company becomes the first commercial contractor to fly United States airmail. Loading the 2-AT plane is Henry Ford himself. Looking back at him is Mrs. Ford, and, to the right of her, are William B. Stout and Edsel Ford, FORD MOTOR COMPANY

in terms of frail World War I biplanes and barnstorming pilots who flirted with death every time they went up. Even before the *Shenandoah* tragedy, the Fords had conceived a dramatic way to communicate to the public the reliability and safety of commercial aviation.

Automobiles had faced the same problem of acceptance a quarter century before. Untried, crude, the first cars were originally thought of as only for daredevils and the very wealthy. But Charles Glidden of Boston, Massachusetts, had come up with a scheme to prove to the average American citizen that cars were for him also. He organized a series of long automobile trips over public highways. The idea was that when people saw the cars close up they would accept them as a practical form of transportation. Most American auto manufacturers, including Ford, had participated in these Glidden Tours. Held for almost a decade, they had done much to popularize auto travel.

Now Henry Ford saw a similar tour of airplanes accomplishing as much for aviation. For years the Detroit Aviation Society had made a bid for the annual national air races to be held in Detroit. In 1925 Henry Ford went to this society's offices with a proposition. Why not this year withhold the bid and, instead, sponsor a national air meet that would promote safety and reliability? The Detroit Aviation Society accepted Ford's proposal, and, soon after, an announcement was made: On September 28, 1925, a National Air Reliability Tour would start from Dearborn with the object of "selling" aviation to the nation by proving that airplanes could carry passengers and freight safely and quickly over long distances.

Any commercial airplane (military planes were excluded) could be flown in the tour, and all the bona fide manufacturers were invited to participate. The only requirements for planes were that they had to have a speed of at least 80 mph and could carry a payload in passengers or freight of 0.5 lbs. per cubic inch of their engine displacements.

The planes in the Reliability Tour would fly approximately nineteen hundred miles in a series of legs. On the first day of the tour, they would fly the first leg from Ford Airport to Fort Wayne, Indiana, and then to Chicago. The next day they would complete the second leg to Moline, Illinois, and

Des Moines, Iowa. The day after that, the third leg to Omaha, Nebraska, Saint Joseph, Missouri, and Kansas City, Missouri. The following day, the fourth leg to Saint Louis, and the day after that, to Indianapolis, Indiana, and Columbus, Ohio. The last and sixth day, the planes would fly to Cleveland and then back to Detroit.

The Reliability Tour was not a race. Thus, each plane that made all the stops and flew safely back to Detroit would be considered a "winner." As such, the name of the returning plane and pilot would be engraved on a huge silver cup designed by Tiffany's in New York.

The Fords went all out to make the Reliability Tour a success. Edsel Ford himself donated the cup, which cost $7,000. The tour would start and end at Ford Airport. While the tour was in progress, an aviation exposition would be held at the airport. As part of this exhibition, Eddie Stinson, the well-known Detroit aviator, would attempt to set a new record for sustained flight, and army flyers from nearby Self ridge Field would stage aerial stunts, "doing everything an airplane can do."

A week before the Reliability Tour was to begin, army flyers surveyed the route to be followed, and in the next few days the planes scheduled to fly in the tour began arriving at Ford Airport. Sixteen planes in all were to fly the route. Included were a Curtiss Carrier Pigeon, two Martins, a Waco, a Swallow, a Fokker, and a Junkers. The Fords were particularly interested in the Fokker and Junkers, since the former plane, although made of wood and veneer, was a trimotor, and the latter, although single-engined, was an all-metal plane with a corrugated skin and cantilevered wing. The Ford entry was a 2-AT. Its pilot was again Edward Hamilton.

Edsel Ford got the Reliability Tour underway when he dropped the starter's flag. At four-minute intervals, the planes raced bravely across the field, taking off and heading for the first stop, Fort Wayne, Indiana, 135 miles away. There, and everywhere else that the planes landed in the next few days, they were enthusiastically welcomed, the tour attracting large crowds and tremendous publicity.

A week later, eleven of the planes, including the Ford 2-AT, made their way safely back to Ford Airport. Only 68.7 percent of the planes returned, and the leisurely route the tour had followed was only slightly more than half the distance between New York and San Francisco, but no one was particularly critical. The Reliability Tour had been a great success, so much so that the Fords and the Detroit Aviation Society decided to hold another one the following year. But this Reliability Tour would have different rules. Already the airplane manufacturers who had participated were beginning to think in terms of competition. They decided the next year's tour would include a point system for grading pilots and planes in order that an actual winner could be determined.

Henry Ford meanwhile got another idea. He had begun thinking about the Kelly Act, and the more he thought about it, the more he leaned to the notion that Ford could participate more fully than just build planes for the air services that would carry the mail. After all, Ford Air Transport Service, the Ford Company's private airline, was the nation's largest and most successful air service. Why shouldn't Ford itself bid on some airmail contracts?

Postmaster General Harry S. New had the same idea, and in September, 1925, he came to Dearborn to confer with Henry Ford about it. New's statements to the press were noncommittal ("because of Ford's growing interest in aeronautics, something profitable for the nation should come out of our conferences"), but privately he urged Ford to bid on two airmail routes—one between Detroit and Cleveland, the other between Detroit and Chicago. "It's logical," New told Ford. "You're flying these routes already."

Ford did make the bids. He offered to carry the mail on these routes for 67.5 percent of the face value of the airmail postage on the letters transported. He also agreed to set aside for the service three 2-ATs, two to be used regularly, one to be kept in reserve.

In January, 1926, when the contract awards were announced, Ford was awarded the two airmail routes he had bid on. The other six airmail routes were given to various air services throughout the nation, many of which eventually grew into America's largest airlines. The route between Seattle and Los Angeles was given to Pacific Air Transport. This company had been founded in 1915 by an ex-stunt flyer named Vern C. Gorst. When his Martin pusher-type plane had cracked up, Gorst, unable to afford a new plane, had gone into the bus-line business. Out of aviation almost a decade, he hankered to return and saw an opportunity when the Kelly Act went into effect. Raising some capital from his bus-line associates, Gorst surveyed the Seattle-Los Angeles route, finding the best ways through the mountains and locating, en route, fairgrounds and pastures that could be used for landing sites. Total expenses for his survey besides the plane rental were $43 for gasoline and $5.25 for oil. When Gorst submitted his bid to the Post Office, it was leery of his company's finances. But Gorst sold his bus line and made arrangements to trade off stock in Pacific Air Transport for planes, pilots, and gasoline. As a result, he got the route.

The route from Elko, Nevada, to Pasco, Washington, went to Varney Airlines, founded by Walter T. Vamey, an ex-World War I pilot. Returning home from the service, Varney had started a flying school at San Mateo, California, and he also operated a small air-express service around San Francisco Bay. Varney bid on the Elko-Pasco route, figuring no one else would want a contract to fly from nowhere to nowhere. He was right. Varney Airlines, the only bidder, was awarded the route.

Another air service to be awarded a route was Western Air Express, a company started by Los Angeles newspaper publisher Harvey Chandler and real estate promoter William M. Garland. They were peeved because the Post Office's main Columbia Route ended in San Francisco rather than in Los Angeles. So they raised $250,000 to form Western Air Express, a company they hoped would put Los Angeles on the aviation map. Their first step was to bid on the Los Angeles-Salt Lake City contract.

Northwest Airways was awarded the Twin Cities-Chicago airmail route; Robertson Aircraft got the route from Saint Louis to Chicago; and Colonial Airlines won the New York-Boston route. The New York-Atlanta route was awarded to Pitcairn Airlines, whose owner, Harold F. Pitcairn, was a small manufacturer of open-cockpit planes and also owned a landing field near Philadelphia. According to the provisions of his bid, he would receive three dollars per pound to fly the mail.

The air service awarded the Kansas City-Dallas route was organized in Stout's office at Ford Airport. This was National Air Transport, a product of the fertile mind of C. M. Keys, an *ex-Wall Street Journal* editor who decided to promote aviation companies instead of writing about them. Eventually

Bringing up the mail for the inaugural airmail flight on February 15, 1926. FORD ARCHIVES PHOTO

establishing the nuclei of what would become two of the United States's largest airlines (United and TWA), Keys—unlike most airline organizers at the time—thought in big terms only. Involved in all kinds of aviation companies (he was a director of a half dozen corporations, including Curtis's Aeroplane and Motor Company), Keys had tried to organize an overnight air-freight service between New York and Chicago'. At various points he tried to involve the Fords, and once persuaded Edsel to become the president of the venture, but Henry Ford forced Edsel to back out, saying his Ford Motor Company duties prevented him from accepting this responsibility.

Keys did get some financial backing from the Fords, and he scored a coup for the new company by signing on Colonel Paul Henderson as general manager. Henderson resigned as Second Assistant Postmaster in charge of the Air Mail Service to take the position. Keys then sought more capital. Working with Carl Fritsche of the Aircraft Development Corporation and helped by the Fords' support, he lined up several Detroit industrialists as backers and then tried to obtain more in Chicago. But a luncheon attended by leading Chicago bankers was a bust; all were railroad-oriented, and no money was raised. Keys was almost ready to quit when a young Chicago associate made an observation. He had noticed that most of the Chicago bankers at the luncheon had sons who had flown in World War I. Why not have another luncheon and invite the sons? Keys took the advice, and within ten minutes after the second luncheon began, the money was raised. Backing his new company would be such men as P. K. Wrigley, Jr., Lester Armour, Philip Swift, and Marshall Field III.

In March, 1925, Keys, Henderson, Fritsche, and some of the bankers met in Stout's office to formally organize National Air Transport and to discuss plans for its operations. Every seat was taken. When Henry Ford arrived late, he sat on a wash bucket turned upside down.

As it turned out, it would be several years before National Air Transport would get its New York-Chicago air service organized, but meanwhile NAT bid on the thousand-mile airmail route from Chicago to Dallas by way of Kansas City, and the Post Office gave it the route.

Once the contracts had been awarded, the air services receiving them busied themselves getting ready to fly the mail. Routes were surveyed again, signals and beacon systems were installed, pilots were hired, and airplanes purchased. As Henderson had predicted when Stout and Mayo had visited him in Washington, D.C., there was no demand for 2-AT planes for this service. It was not just that pilots preferred smaller, lighter, and more maneuverable planes. It was also that none of the air services could afford the $20,000-$25,000 cost of the 2-AT. All they needed were planes to fly the mail. Bigger planes like the 2-AT, that could carry passengers as well, were wasteful in their eyes. As a result, the planes purchased by the various air services preparing to carry the mail were almost every type but Ford 2-ATs. Robertson stuck to DH-4s. Varney ordered six small Swallows. Pitcairn bought one of its own planes, the Pitcairn Mailwing 5. And National Air Transport purchased ten Curtiss Carrier Pigeons and thirty-five government surplus Liberty engines.

Even so, Ford began to sell 2-ATs to outside customers. The first plane was sold to John Wanamaker Stores, which purchased it for use as a company plane to transport executives and freight between its New

York and Philadelphia stores. Wanamaker's announced it had ordered nine similar 2-ATs and would eventually operate a company airline that would extend as far south as its stores in Florida.

Henry Ford was particularly pleased at this sale, since Wanamaker's had been the first store of any kind to sell a Model T Ford. To publicize the 2-AT sale, arrangements were made to fly the plane to New York's Mitchel Field in October, 1925, when the National Air Races would be held there.

The races attracted the largest assemblage of commercial and military aircraft ever seen up to that time. One of the show's highlights was the arrival of the Wanamaker 2-AT. A large crowd was on hand when the big corrugated-metal plane roared out of the west and landed on the field. Greeting pilot Leroy Manning when he stepped out were Edsel Ford and Grover Whalen, Wanamaker's president. Manning handed to Whalen a personal letter from Henry Ford. "We take great pleasure in knowing that the first Ford-built plane to be delivered to a customer will be going to John Wanamaker," it said. "The plane will be kept at Mitchel Field during the Air Meet and then put on display in Wanamaker's New York store."

Other 2-AT sales followed, including sales to airlines. In December, 1925, Mayo announced that both Western Air Express and National Air Transport had placed orders. Soon after, four more 2-ATs were sold to Florida Airways. This company had been organized to carry passengers and freight between Miami, Fort Myers, and Tampa. Its founder was Reed M. Chambers, a World War I flying ace who had commanded the 94th Aero Squadron, for which Eddie Rickenbacker had flown. In organizing Florida Airways, Chambers had signed on former army pilots; but financing was a problem. The 2-ATs the company purchased were used models flown by Ford Air Transport Service. The arrangement was that when the private Ford airline got its new trimotor planes, they would send four of the 2-ATs they had been using down to Florida Airways.

The first airline to get its airmail operations going was Ford. On February 15, 1926, the Ford Motor Company became the first commercial contractor to fly airmail in the history of the United States Post Office. Ford Airport was covered with snow. Besides the Fords, Stout, and Mayo, many notables were present. Making a speech over a national radio hookup, Henry Ford said, "The pioneering in plane building and operation is past. It now remains for men of business to take hold of the opportunity."

Captain Lawrence G. "Larry" Fritz climbed into the 2-AT flying to Chicago, and Captain Russell Kirkpatrick entered the cabin of the 2-AT headed for Cleveland. While they prepared to take off, a Model T pickup truck arrived with the mail. Photos were taken, mail sacks loaded aboard, and the planes began taxiing down the field. When they were no longer in sight, Henry Ford revealed to the press that on the Cleveland plane was a letter he had sent to his good friend, Thomas A. Edison. "I think this is a great step forward," Ford had written.

The other airlines started their airmail service later. In April, 1926, Varney, which eventually would form part of United Airlines and have over a million square miles of air routes in the Pacific Northwest, began flying the Elko-Pasco route. The same month, Western Air Express, which eventually would become an integral part of TWA, inaugurated service on its Salt Lake City-Los Angeles route. Western added a feature that none of the other mail-carrying air services bothered with. It announced it would also take passengers. No more than two could be carried each flight, and they would have to sit on folding chairs put down between the mail sacks. But it *was* passenger service.

In the same month, Robertson Aircraft began flying the Saint Louis-Chicago route. The pilot it chose for its inaugural mail flight was a tall, lanky youth with light curly hair and a boyish face, and from the beginning, flying had been in his blood. After attending the University of Wisconsin engineering school for only three semesters, he left to enroll in a flying school. Then he became a flying cadet at Kelly Field in Texas and, later, a captain in the Army Air Corps Reserve. Until he joined Robertson, he had been flying the mail for the United States Post Office. His name was Charles Augustus Lindbergh.

Pacific Air Transport did not get its Seattle-Los Angeles airmail operation started until the fall of 1926. Vern Gorst had done yeoman work in preparation. He had persuaded towns along the route to build community airstrips. He had gotten Standard Oil to paint in huge letters on their gasoline storage tanks and warehouses the names of the cities in which they were located. He had installed makeshift beacons on high buildings and poles. For his service, Gorst had put together a motley fleet of Ryan monoplanes, Waco Swallows, and Wright Travel Airs. To fly them, he had hired ten ex-army and ex-Post Office pilots. But the going was rough. In the first year of operation, Pacific Air Transport lost not only money, but three of its pilots, who were killed in crashes.

Meanwhile the Fords kept busy with all their aviation operations. They were continually improving their facilities. On their Chicago run, for example, they found smoke and fog were often so heavy over the city that the landings at Chicago's Maywood Field were becoming difficult and even dangerous. So they purchased fourteen hundred acres of land at Maynard, Indiana, twenty miles southeast of Chicago, and here they built their own airfield. Similarly, at Cleveland, to keep up with increased activity at that city's airport, they erected new hangars.

And the Fords kept exploring new opportunities. Hoping to extend their own private airline to the east, they began talking to Colonial Air Transport, which had the New York-Boston airmail route. From these talks came a proposal for the two companies to organize a new service that would link up Detroit with Boston, New Haven, Buffalo, and New York.

While the Fords considered this projected airline, Stout was rapidly finishing up work on the new three-engined Ford plane. Henry Ford already had thought of a dramatic way to introduce the plane—if it could be completed in time. A Ford plane had already been talked about as possibly being used on a transpolar flight to be attempted by the Arctic explorer Stefansson. But actually the Fords favored another Arctic explorer's plan to fly over the North Pole. This was Richard Evelyn Byrd, a prominent navy officer just returned from the Arctic.

Scion of a famous Virginia family, Byrd was a direct descendant of Lord Delaware and of William Byrd II, the founder of Richmond. At the age of thirteen, he completed a trip around the world alone, and in 1908 he entered the United States Naval Academy, where he was an outstanding athlete despite his short stature. Byrd took part in the capture of Vera Cruz in 1914, but four years later he was forced to retire from the navy when he injured his leg. Returned to duty during World War I, Byrd was assigned to Pensacola Naval Air Base in Florida. Here he learned flying and miraculously survived a head-on airplane collision. The rest of the war he spent in Canada, heading up United States naval forces there.

After the war, Byrd became prominent in aviation. In 1919 he helped plan the navigation of the three big Navy NC seaplanes that made the first transatlantic flight. Later, he drafted and successfully promoted through Congress the bill that created the Navy Bureau of Aeronautics. He also established the Naval Air Force Reserve.

In 1925 Byrd began planning his first expedition to the Arctic, but, learning that Donald MacMillan was preparing a similar expedition under navy auspices, Byrd decided to join forces. The MacMillan expedition had two Loening amphibian planes, and in these planes Byrd made two unsuccessful attempts to fly over the North Pole. Byrd reasoned later that these flights had been made at the wrong time of year. If they could have been attempted in May, he felt he would have succeeded.

Around this timetable, Byrd began planning a new expedition that would attempt a transpolar flight and also explore parts of northern Canada, Alaska, and Greenland. Byrd's plan was to sail to Spitsbergen in March, 1926, and then fly to Cape Morris Jesup, four hundred miles from the Pole, where a base would be established. From here he would attempt the polar flight.

By the end of 1925 most of the expedition's planning had been completed. The pilot for the flight would be Floyd Bennett, a navy noncommissioned officer who had been Byrd's pilot on the MacMillan expedition. Bennett had gotten a leave of absence from the navy and was ready to go.

Byrd needed more financing. He himself had borrowed $20,000 to put toward the expedition, but an additional $100,000 was required. Byrd started making the rounds of wealthy Americans, seeking five investors to take out $20,000 shares. He promised to pay them back from the earnings he expected to receive, after his return, from lecturing and writing.

Despite the fact that the stock market was soaring to unexpected heights, Byrd had trouble raising the funds. But then he contacted Henry and Edsel Ford, and his problem was solved. The Fords decided to help him in every possible way.

They told him of the new three-engined plane being rushed to completion, and offered it to him for the flight over the North Pole. They took out a $20,000 share in the expedition, and Edsel himself found the remaining shareholders. Vincent Astor was one; another was John D. Rockefeller, Jr. Rockefeller admitted to the press that he had never done anything like this before, but had been won over by a telephone call from Edsel.

Stout standing in front of the ill-fated 3-A T. FORD ARCHIVES PHOTO

In November, 1925, the Fords were notified by Stout that the three-engined version of the 2-AT would be completed by the end of the month. In eager anticipation they awaited Thanksgiving Day, the date selected for testing the new plane. When the 3-AT (the model designation of the plane) was rolled out on the runway, it did not resemble the 2-AT, even though it had been built on the same type of structure. Compared to the 2-AT, the 3-AT trimotor plane looked top-heavy and awkward. The position of the center engine made it seem like an afterthought. It had been located in the nose—not in the center, but on the bottom, and the fuselage over it had been rounded off in such a manner that the nose was gondola-shaped, making the body of the plane appear more like a dirigible than an airplane. In the gondolalike nose, high over the center engine where the pilot's cockpit was located, and in front of the pilot, was a window that one person described as a "misplaced bay window."

Despite the plane's awkward appearance, Stout had high hopes for its performance. The three new Wright J-4 Whirlwind engines gave the 3-AT a total of 600 hp, 180 hp more than the one Liberty engine on the 2-AT. Also, many technical changes had been made. To increase the streamlining, the starboard and port engines had been installed right in the wings, instead of below them. Also, the wing of the 3-AT had been widened and extended, and its landing gear strengthened to bear the weight of the extra engines.

All the leading Ford executives, including Henry and Edsel, showed up that crisp day to see the 3-AT fly. The test pilot was Major R. W. "Shorty" Schroeder, the famous army aviator who held the high altitude record. He had recently joined the Ford Motor Company as one of the company's chief pilots.

Schroeder climbed into the plane and took his seat in the gondolalike pilot's compartment. The three engines were started, and after they had warmed up, he gunned them and started down the runway. The big corrugated-metal trimotor took off easily and, its engines roaring lustily, started climbing into the sky. Everyone on the ground began to smile. Then, suddenly, Schroeder was seen turning around and coming back to land. As he approached the runway, the 3-AT began acting erratically. It dropped down thirty feet, then bounced up thirty feet, and kept going up and down like a Yo-Yo. Managing to get the big plane back into the air, Schroeder circled the field again, and then landed—but only by using full power.

Those watching on the ground were horrified. Somehow by altering the configuration of the 2-AT—by changing the shape of the nose and installing the center engine down below—Stout must have altered the plane's flight characteristics.

Schroeder confirmed this view. He told the group after he landed that the 3-AT should be able to fly at 110 mph, but its actual flying speed was less than half of this. Even worse, the only way the plane could land was to come in at the unsafe speed of 90 mph. "My advice is to forget this plane," he said.

The Fords and Stout could hardly believe it. Another pilot was asked to take up the 3-AT in the hope that the plane might perform differently. But this test flight was similar to Schroeder's. The 3-AT was a bust. It could not carry enough payload. It was too slow. It was not a safe plane to land. It could not meet Henry Ford's criterion of being able to land on one engine and keep flying with two.

Henry Ford flew into one of his characteristic rages. For months he had been quoted in the press about this new plane and how it would revolutionize commercial aviation. He had also offered the plane to Byrd. And now—it could hardly fly!

"This plane is a mechanical monstrosity and an aerodynamic absurdity." He turned to Mayo. "From now on keep Stout out of the design room. I bought a lemon and I don't want the world to know it."

Stout thought Ford would cool down in a few days, but the auto magnate did not. Stout tried to point out to Mayo and Ford that the lack of sophistication in aviation design then made it impossible to predict a plane's performance until it was actually built. He also emphasized that in his opinion the 3-AT was aerodynamically sounder than the 2-AT, and that with some modifications it would perform satisfactorily. But Ford would not listen. He did not intend to sever his connection with Stout, but he wanted a new man in charge of aviation design—a man who would build him a trimotor that would fly.

Selected to replace Stout as head of the airplane division was Harold Hicks, who had worked on the Liberty engine in World War I and later joined Ford as a special design engineer. Hicks had designed an experimental motorboat for Edsel, and had helped Henry with a projected gasoline-powered streetcar.

David A. Weiss

Stinson Detroiter. It was manufactured by Eddie Stinson, well-known Detroit pilot who had flown the Maiden Detroit *in its initial flights. While the 3-AT was under construction, Stinson announced a trimotor version.* PICTURE COLLECTION, NEW YORK PUBLIC LIBRARY

Stout was destined to stay on at Ford for several more years, and he would make many more valuable contributions—but not in design. This was now Hicks's bailiwick. The decision to remove Stout from design was, as historian Allan Nevins pointed out, "very unfair since Stout's work in metals and aerodynamics formed the basis of Ford's entire activity in aviation." Yet, as Nevins also pointed out, Hicks's closer relationship to the Fords gave him a capability of "being better able to develop manufacturing within the Ford organization." The changing of the guard was accomplished smoothly. Stout was given the assignment to tour the country to promote aviation, and he was also encouraged to start organizing the air services company which he had formed with his ex-stockholders at the time Ford Motor Company had taken over the manufacturing operations.

Hicks's assumption of his new duties was dramatically abetted by a mysterious fire that destroyed not only the ill-fated 3-AT, but also the airplane factory in which all Stout's work had been done. The Fire Department was able to save the hangar, but the factory burned to the ground, along with the 3-AT and four almost-completed 2-ATs being built for the Ford Air Transport Service's fleet. All that was left of the planes were various steel parts, still eerily in position in the ashes on the factory floor. Destroyed also was a stockpile of engines as well as Stout's jigs and tools.

The monetary loss was estimated at $200,000, and even more damaging was the psychological setback of having to halt 2-AT production just as outside sales were beginning to be made. Stout took the fire harder than Ford. He was gloomily poking around the charred debris when Ford came over.

"Don't look so sad, Stout," Ford said. "It's the best thing that could have happened. Now we can build the kind of factory we should have built in the first place."

Hicks took advantage of the opportunity to reorganize the division. Called now the Airplane Manufacturing Division instead of the Stout Metal Airplane Division, it moved into temporary quarters in the garage of the Ford Motor Company's engineering laboratory. The space was too small for manufacturing, but experimental work could be carried out there. Meanwhile, a crash program was started to build a new airplane factory near the old one. It would be three times the size of the previous factory, and near it would also be constructed a hangar with a capacity of fifteen planes.

Since Stout was now out as a designer, and Prudden had been discharged earlier, the only Stout-trained design engineer remaining was Tom Towle. Hicks increasingly relied on Towle, and, to provide additional design backup, three Massachusetts Institute of Technology graduates were added to the staff: James McDonnell, Otto Koppen, and John Lee. McDonnell, later to found the McDonnell Aircraft Corporation, had once been a member of one of Stout's model-airplane clubs.

Despite the changeover and the fire, no time was lost getting a new three-engined plane underway. Hicks directed Towle to bring what blueprints were available to the new temporary factory, and designing began immediately on the new plane.

The Fords, annoyed at the delay in producing a three-engined plane, were more convinced than ever that trimotors were the answer to commercial aviation's future. Others in aviation were also coming around to this opinion. National Air Transport, Keys's company, had recently made a survey of European commercial aviation. It revealed that multi-motored planes were becoming popular there. Shortly after, a group of American aviation leaders issued a statement that if commercial aviation was feasible at all, a multiengine transport was essential.

In the United States, Fokker had for several years been the only important manufacturer of trimotor planes. But others were about to introduce models. The Detroit pilot Eddie Stinson had entered manufacturing, and he announced a new plane, the Stinson Detroiter, which would have three engines. This announcement spurred on the activity of the Aircraft Manufacturing Division to build its new trimotor. Once it did, Ford could take over most of the market, the company felt, since its plane would be all-metal, while both the Fokker and Stinson were made of wood and veneer.

But the weeks flew by, and Hicks and Towle—working around the clock—had not completed the new trimotor plane. Stout meanwhile was busy in his new role of promoting aviation. On the first anniversary of the Ford Air Transport Service, it was he who arranged for appropriate publicity. Holding a press conference, Stout announced that the performance of the planes and the pilots had been outstanding the previous year. "This entire twelve months of flying has been accomplished without injury to anyone," he said, "and with a remarkable freedom from mechanical trouble."

Stout revealed that the Ford Motor Company's private airline had made more than 1,000 flights, and flown more than 26,000 miles at speeds close to 100 mph. From 1,000 pounds to 1,500 pounds of freight had been carried on each flight, and maintenance on the planes had been so simple that Ford had been able to operate the service on only four planes. "With these four planes, we were able to fly all our schedules, which included three trips daily on the Chicago run and two trips daily on the Cleveland run," he said.

Stout added more details to emphasize the reliability of the service. One plane on the Chicago run had racked up 1,000 hours of flying time. Another plane had once transported a complete Model T automobile to Cleveland. All in all, 963 of 979 scheduled flights, or 98.3 percent, had been completed. "All this has been accomplished without any blowing of trumpets," he said, "but just as an everyday, routine transportation proposition."

Stout also at this time began to organize his Stout Air Services Company. When he had first formed this company, his intention had been to establish an airline along the principles he had outlined to Henry Ford that day when they had sat outside the factory and discussed the future of aviation. Stout had said then that he thought the big profits were in the air service area, and not in manufacturing. In the intervening months he had become even more convinced of this.

Before Stout had become so involved in the ill-fated 3-AT plane, he had given considerable thought to his plans for a Stout Air Services airline. Somewhat grandiosely he saw this airline as linking all the nation's largest cities. He had even worked out on a map of the United States a series of interconnecting routes. But the realization soon came that the capitalization of Stout Air Services—it was only $350,000—was insufficient *to* support such an operation.

Too busy with the 3-AT to devote much time to his Stout Air Services Company, Stout then got the idea that perhaps he should invest some of the company's money in other airlines. For a time he followed this approach, his company purchasing stock in both Western Air Express and Northwest Airways. But now, with more time on his hands, he decided to return to his original concept and begin organizing his own airline.

Realizing he would be unable to finance a very large airline, Stout thought that if he could operate a small airline along the principles he had developed for passenger travel, he could influence the other airlines and upgrade passenger service throughout the nation.

As Stout saw it, if the right kind of passenger service was available, an airline could succeed even if it did not carry airmail. Such a view was not shared by any airline then operating. To them, passenger service was definitely subordinate to mail service. The few passengers carried usually sat between mail sacks, and the airline had the right to "bump" passengers at any landing field if it wanted to take on more mail.

Why anyone would want to be an airline passenger was hard to understand. Besides the possibility of being dropped off miles from his destination, a passenger more often than not had to fly in an open cockpit. This meant he had to wear the same kind of flying outfit as that worn by the pilot. Before boarding the plane, he was handed coveralls, a flying helmet, and goggles. He was also given a parachute, but the pilot seldom explained how it operated. It was small wonder that Northwest Airways, which flew from the Twin Cities *to* Chicago, carried only 106 passengers in all of 1927.

A few airlines were becoming more interested in passenger service, and had begun in a small way to promote special passenger flights. Universal Aviation Corporation, which later merged with Robertson Aircraft, had launched a service between Saint Louis and Chicago that carried six passengers on each flight. The planes used were Fokker trimotors, which had one compartment for mail and freight, and another which served as a passenger cabin. Because the cabin was enclosed, no flying outfits were needed.

Stout, however, saw passenger service as far more accommodating and comfortable. He saw no reason that air travel could not be put on a par with travel by railroad Pullman. With this thought in mind, he decided that Stout Air Services would pioneer a new type of airline where the passenger would reign supreme. He decided to start on a very small scale, operating first an experimental service between Detroit and Grand Rapids. By automobile, the trip took four hours. By air, Stout figured it would take an hour. In the summer of 1926, Stout and Stanley Knauss surveyed the route. On August 1, 1926, operations began with a 2-AT. In the next few years Stout Air Services would become the best-known passenger airline in the United States and the only regularly scheduled airline to survive the vicissitudes of its early years.

Although the Fords were closely watching the progress on Stout's airline, they were far more interested in how Hicks and Towle were doing with the new plane. By May, 1926, final work was being done on the plane, and preliminary tests were made. On June 11—little more than six months after the factory had burned down—the new Ford airplane was ready for its first official test. The two Fords, Mayo, and Stout watched the big trimotor being wheeled out onto the runway.

In building the new 4-AT (the plane's model number), Hicks and Towle had basically followed the approach used by Stout in constructing the ill-fated 3-AT, namely, they had taken the single-engined 2-AT, enlarged it somewhat, and added two engines. But the 4-AT was a vast improvement in appearance over the 3-AT. The lines flowing back from the plane's nose were neat and clean, and the nose itself was not gondola-shaped, but tapered and trim. The center engine was not positioned grotesquely low in the gondola, but was installed symmetrically in the center of the plane's nose.

Like the 2-AT and the 3-AT that preceded it, the new Ford Tri-Motor had an oval-shaped door, a sharp-pointed tail, and that familiar corrugated-metal skin. Unlike the earlier models, however, it had a wing curve of greater lift, its control wires were on the inside, and it was bigger. With a wing span of 74 feet and a fuselage that was 49 feet 10 inches long, it was the largest commercial land airplane ever built in the United States up to that time.

"Shorty" Schroeder was again the test pilot, and he lost no time getting the new Ford Tri-Motor into the air. From the moment he taxied down the runway, the plane looked like a winner. In only seven seconds it was airborne, and in the next hour, while Schroeder put the plane through its paces, the 4-AT flew perfectly. When Schroeder finally landed and stepped out, he rushed up to Henry Ford. "This plane's got what it takes."

Chapter Six
The Tin Goose Roars

Everyone had been so sure of the new Ford Tri-Motor's success that another demonstration had been planned that very afternoon for representatives of major United States airlines. Represented were National Air Transport, Western Air Express, Colonial Air Transport, and Florida Airways. Briefings had been held earlier by Hicks, who explained to them that this new Ford plane would be able to meet all of their transport needs. After the representatives saw the 4-AT fly, and went up for rides themselves, they needed little convincing.

Returning to their company headquarters, they talked excitedly about the new Ford Tri-Motor and its performance. This new plane really had power, they said. Just one look at its big, thick wing and you got the feeling it could lift anything. And indeed it almost could, since, incredibly for the day, its payload was 1,725 pounds. The new plane was fast, too, the result of not only its three new 200 hp Wright J-4 engines but also its streamlining. There were no exposed struts or wires. Even the engines, which were installed below the wings instead of in them, had been placed in nacelles, or pods, which reduced the drag. As a result, the new Ford Tri-Motor had a maximum speed of 120 mph and a cruising speed of 107 mph. No other commercial plane could top this.

The new plane was also tough and rugged. All-metal construction had been used throughout. The wing framework had been constructed of three panels, each of which had been built up of duralumin spar beams to form a truss-type beam. The wing ribs had been similarly constructed. The fuselage framework had been constructed of duralumin open-channel sections also riveted together to form a truss-type structure. For additional strength, the plane's corrugated-metal skin was riveted to the fuselage framework.

Preceding Page: The 4-AT—first of the famous Ford Trimotors. FORD ARCHIVES PHOTO

The skin was made of Alclad, a metal sandwich with a duralumin sheet in the middle, covered by layers of pure aluminum on both sides. This sandwich construction made the skin corrosion resistant even at the edges and rivet holes. For additional corrosion resistance, all exposed parts of the duralumin fuselage and wings were treated with a protective lacquer coating that was both moisture proof and rust-resisting.

The big 4-AT could also fly well. Even fully loaded, it could take off in just a few plane lengths, and it could climb 900 feet per minute to an altitude of 15,000 feet.

In designing the big new Ford Tri-Motor, great attention had been paid to the cockpit and the controls. For extra safety, there were dual controls: the copilot as well as the pilot had his own steering wheel. Visibility was almost 100 percent. Sitting in bucket seats before the controls, the pilots had on two sides large windows that could be opened. In front was an even larger window serving as the windshield.

The dashboard of the new plane was simplicity itself. It had no more than a dozen instruments and gauges, but all essential ones had been included: airspeed indicator, magnetic compass, clock, altimeter, rate-of-climb indicator, turn-and-bank indicator, fuel gauge, tachometer, oil-pressure gauge, and oil-temperature gauge. Below the dashboard was a cluster of three knobs, which controlled the power to the engines. Near these were the engine switches, one for each engine.

All of the controls in the new 4-AT could be handled easily. The pilot could change the plane's trim by turning a small crank over his head. The ailerons (these were of the offset-hinge type) were controlled by a wooden-spoked automobile steering wheel, the same as used on the deluxe Model T car. This wheel was attached to the top of an upright column, and when it was turned to the right or left, the ailerons wiggled in such a way that the plane turned to the right or left. To lower the nose of the plane to bring it down, the pilot raised the elevators in the tail by pushing the wheel forward. To raise the nose to climb, he pulled the wheel back. Between the two pilot seats was the plane's brake, a simple gearshift lever, also taken from the Model T car. The 4-AT's landing gear was of an advanced type with a split axle and rubber shock absorbers. Steering on the ground was accomplished with the plane's engines.

Maintenance of the new Ford Tri-Motor had been simplified. The plane's gas tanks, which held 271 gallons of gasoline, were easily accessible in the wings. The engines were also accessible. As for the corrugated skin, it seldom needed to be patched, and it could be cleaned by simply hosing it down with water.

The new Ford Tri-Motor was also safe, or as safe as an airplane could be at the time. Its construction, its power, its three engines—all added up to more reliability than those of any other plane. And just as Henry Ford had wanted, the plane could still fly easily even if one motor failed, and if both engines failed, it still had enough power to land or to get into the air.

No plane of the day emphasized passenger comfort as much as the new Ford Tri-Motor. The passenger compartment back of the pilot's compartment (the two were separated by a bulkhead with a door) contained more than 450 cubic feet. With a length of 16 feet 3 inches, a width of 4 feet 6 inches, and a height of 6 feet, it had the most spacious passenger cabin of any plane then being produced in the United States.

The 4-AT seated eleven passengers. There were two rows of seats running down the length of the passenger cabin, one row with five single seats (and the door), the other with six single seats. In between was a narrow aisle. The seats themselves were of wicker, set in aluminum-alloy frames. The way they were positioned, each passenger had a large half-moon window of shatterproof glass to look through.

The first 4-AT, although a sensation for its day, was admittedly crude compared to modern jetliners. The plane had no heater, for example. As a result, passengers would later call it a "flying icebox." Also, it was not particularly attractive, some calling it from the beginning a "flying washboard." The noise and vibrations from its engines were overwhelming. To be heard in the pilots' compartment, pilots had to shout. In the passenger compartment, despite soundproofing and insulators, the noise was, as one passenger described it, "like a hundred gremlins hammering on a barn door with little hammers." (To reduce the noise, cotton was given to passengers to stuff into their ears. At best its effect was more psychological than acoustical.)

Despite these limitations, the Ford Tri-Motor represented a tremendous technological advance, and the airlines realized it. They liked everything they saw about this plane: its power, its safety, its ruggedness, its capacity, even its price. This, Ford had set at $50,000 f.o.b. Detroit, and when a customer came to Ford Airport to pick up the 4-AT, it was ready to fly—so ready that Ford, as a public relations gesture, even had the plane's gasoline tanks completely filled.

Just as the Fords had reasoned, they had no real trimotor competition. The 4-AT's size, power, passenger capacity, safety, all-metal construction, and cantilevered wing put it into a class by itself. The Stinson Detroiter trimotor was much smaller, and did not have all-metal construction. It did have some steel tubing in its framework, but it also had wooden strips, and its framework was covered with impregnated canvas. As for its wings, they were not cantilevered but were supported by braces fastened to the fuselage bottom. The plane also had only one engine. This gave it a payload of not more than 1,000 pounds and a passenger capacity of only five including the pilot.

The Fokker trimotor was more nearly like the Ford. It had the same type of configuration; its engines were Wright J-6s; and it also featured dual pilot controls. But the Fokker trimotor was constructed of wood and veneer, and although it looked very much like the Ford, it was very slightly smaller in size.

Fokker claimed later that the Ford Tri-Motor was a carbon copy of his trimotor. He said that Ford had gotten hold of a Fokker trimotor and, within minutes, had engineers armed with micrometers swarming all over it, measuring its contours and dimensions. He even repeated a rumor that Henry Ford had called Hicks in, showed him a photograph of a Fokker trimotor, and directed, "build our new trimotor like that, but make it in metal."

Actually, there had been Fokker trimotors at Dearborn. One had flown in the Reliability Tour the year before. Also, Byrd, when unable to use a Ford plane for his transpolar flight, had purchased a Fokker trimotor. The Fords had paid for the plane, and Byrd in gratitude had named it the *Josephine Ford,* after Edsel's daughter. With this trimotor, Byrd and Floyd Bennett, on May 9, 1926, had succeeded in their flight over the North Pole. They flew more than 1,500 miles in 15-1/2 hours, and each received the Congressional Medal of Honor.

When Byrd returned to the United States, he brought the *Josephine Ford* to Dearborn, and both Fords came to look the airplane over. Even so, the Ford Tri-Motor was not the Fokker trimotor produced in metal. Its design was original, and so was its metal construction. Its wing too was quite different from the Fokker's, having been adapted from National Advisory Committee for Aeronautics contours. And many features introduced by Stout in the ill-fated 3-AT—features such as the thick corrugated skin, vertical fin and rudder, and aft cabin section—had been retained in the 4-AT.

The Fords realized they had a promotable product, and they lost no time launching an extensive advertising and publicity campaign to introduce it. Calling their new airplane "The Ford Superplane," they announced it would make its first public appearance during the second annual Reliability Tour.

On Aug. 7, 1926, this tour would start from Ford Airport, and it would cover about 2,200 miles—a longer distance than the first tour. Flying in it would be twenty-four planes. Walter H. Beech would be flying a Wright-powered Travel Air; Eddie Stinson, one of his own Stinson Detroiters; and Louis Meister, a Buhl-Verville Airster. Also participating was the popular aviator Charles S. "Casey" Jones. Flying a seven-year-old Curtiss Oriole, Jones did not expect to win. He was flying as an extra treat for local audiences. Other planes in the tour included a Ryan, a Pitcairn, and an Alexander "Eaglerock."

The planes took off from Ford Airport on schedule and, after a brief stopover at Kalamazoo, headed for Chicago, the last stop on the first day's leg. A crowd of more than five thousand greeted them at Chicago's Maywood Field, and when the new Ford Tri-Motor landed, cheering spectators rushed up to get a closer look. The next day the planes flew to Hamilton Field in Milwaukee, where a three-day aerial exposition was in progress. By August 11, they had reached the fifth leg of the tour, Des Moines, Iowa, but two of the starting twenty-four planes had dropped out. And so it went until August 21, when those planes still in the tour came down out of the air to land at Ford Airport again.

The first plane to cross the white finish circle painted on the field was the Wright Travel Air, with Walter Beech as pilot. To him Edsel presented the big silver trophy and the lion's share of the $20,000 prize money. Second prize went to Louis Meister in the Buhl-Verville Airster. The plane everyone was waiting for—the Ford Tri-Motor—failed to finish. Neither engine failure nor a structural defect was to blame. En route, an unknown object had struck one of the trimotor's propellers, breaking off a piece. The engine started vibrating so violently that, before anyone could do anything, it tore loose from its mountings with the plane unable to maintain an even keel, the pilot was forced to land.

It was a disappointing blow to the Fords, but it did not affect the sales of the new plane. Even though only one 4-AT had been built and the new factory had not yet been completed, the Fords had no trouble getting orders

After a new propeller and engine were installed in the 4-AT it was purchased by the Ford Air Transport Service and put on the Detroit-Chicago run. The second 4-AT, still to be built, was ordered by Colonel Paul Henderson for National Air Transport. He had seen the big trimotor fly that first day at Ford Airport and had contracted for one on the spot. While the plane was being constructed, a hassle developed between several NAT executives and the Ford Company. Although NAT pilots were looking forward to flying the new Ford Tri-Motor, they did not like the idea of an enclosed cockpit. As they saw it, a pilot had to be outside in direct contact with the elements, with the wind and rain in his face. Stout and other Ford executives tried to argue with NAT, pointing out the advantages of having pilots protected inside the plane. But NAT was adamant; and the windows of the plane's pilot compartment were removed.

By November, 1926, NAT's new open-cockpit plane was ready, and a test flight for it was arranged over NAT's Kansas City-Dallas route. The weather was bitter cold and windy, and since the plane had no heater, both pilots and passengers became chilled soon after the Ford Tri-Motor took off. The pilots in particular had a hard time, since the wind made it impossible for them to keep their woolen face mufflers in place. When everyone returned to Detroit, Stout asked NAT if it wanted to reconsider enclosing the pilots' compartment. NAT did. When the plane was put into actual service in January, 1927, the windows in the pilots' compartment had been reinstalled.

A few weeks later the new enlarged airplane factory was completed. Announcing this fact, Henry Ford said the company was now fully committed to trimotor planes. In the new factory no single-engined 2-ATs would be manufactured; from now on the Fords would produce only 4-ATs. Ford also announced that the new Ford Tri-Motor would be manufactured on a mass-production basis. Such news made the rest of the aviation industry shake its heads. Up to then, no aircraft manufacturer had adopted assembly-line procedures in producing planes. Anthony Fokker shortly before had flatly said there was not enough demand in the United States for planes of any kind to warrant large-scale manufacturing operations.

But the Fords obviously believed otherwise. They proudly called attention to the 60,000 square feet of floor space in their new factory, and said they "had laid it out expressly to accommodate the Ford system of progressive production." The new factory had been built with two straight production lines. At the beginning of operations, each line could produce a plane every two weeks. Later it was hoped each line would be able to turn out as many as two planes a week. These production lines were similar to those used in manufacturing Model Ts. Raw materials came in the back door of the factory. Out the front door rolled the completed Ford Tri-Motor. So complete was the plane that no sooner did it leave the factory than it was taxied over to the airfield runway and test-flown immediately.

Many production techniques used at the airplane factory were new to aviation. Among these were the use of interchangeable parts and the employment of jigs of the same dimension. Another innovation was the use of materials like Alclad, an alloy that combined the corrosion resistance of pure aluminum with the strength of duralumin.

Before the factory was in operation a month, the Fords had orders for fifteen planes. Every possible type of purchaser was represented. The United States Navy, for example, ordered the fourth plane for use as a personnel transport. The army also purchased one for use as a transport.

Various corporations also ordered planes. The first of these was Standard Oil Company of Indiana. Its Ford Tri-Motor, the sixth to be produced, was the first executive airplane ever to fly in the United States. Another Ford Tri-Motor was purchased by Texaco Company, which hired Frank M. Hawks a noted aviator of the day, to be its pilot. Royal Typewriter Company also bought a Ford Tri-Motor, and this plane was converted into "The Royal Air Truck." Built with a special fuselage hatch instead of windows, the plane had space for 210 Royal typewriters. Royal used it for several months to promote its typewriters, flying over fields and dropping the typewriters down by parachute to demonstrate their ruggedness

As a sales gimmick, Ford established a flying school to train pilots to fly the new planes. Actually, the company had an ulterior reason for maintaining the school. It did not want the planes flown by incompetent pilots who might crack them up and give them a bad reputation. "Purchasers of planes are welcome to send their own men to our school for training" the announcement of the school said. "But we must ask them to consider final our decision of their fitness. So important do we regard this provision, we reserve the right to decline to deliver a Ford plane unless the pilot who flies it meets the approval of the officials of our training school."

5-AT Ford Trimotors being produced on a mass-production basis. FORD ARCHIVES PHOTO

The Fords expected no trouble selling the new planes. Once production was accelerated, they intended to use their ten thousand auto dealers as sales agents for the trimotor. One sales agent they even put into the airline business. He was Jack L. Maddux, a Lincoln dealer in Los Angeles.

It was Mayo who suggested to Maddux that he purchase one of the new Ford Tri-Motors and use it to start up a passenger-freight airline linking Los Angeles to San Francisco and San Diego. Not really enthusiastic, Maddux looked for a way out. Talking to some army pilots, he was told the country along the California coast was too rough for scheduled flights. The Ford Tri-Motor would never be able to carry passengers safely over the mountains, they said. When Maddux relayed this information to Mayo, the Ford executive said the army pilots were mistaken. "The planes will perform satisfactorily," he said.

In self-defense, Maddux answered, "O.K., you get into a plane and fly it to me from Detroit to San Diego, and I'll start the airline."

Mayo took up the dare. He assigned Larry Fritz to do the piloting, and as soon as a new Ford Tri-Motor became available, he gave Fritz his last-minute instructions. Taking a map of the United States, he drew a straight line from Detroit to San Diego, and told Fritz, "Follow that route."

Making the flight along with Mayo were several Ford executives as well as selected passengers. It was pioneer flying all the way, but Fritz made it without any trouble. The plane made many stops en route, and everywhere it landed, large crowds turned up at the airfields to see it. At San Diego, the mayor was on hand, along with Maddux and various other city officials. Fritz and Mayo were greeted warmly, and as soon as Fritz refueled the plane he took them up on sightseeing flights over the city.

Maddux took over the Ford Tri-Motor and, true to his word, prepared to open his airline. With this first plane he surveyed the route; then he ordered four more Ford Tri-Motors to serve as the fleet for the new airline. Within a few years, Maddux, 16 Ford Tri-Motors, and his airline was one of the nation's biggest. Later it would merge into TWA.

A Stout Air Services Ford Tri-Motor. Among the many firsts pioneered by the Stout airline were hot meals, flight escorts, and uniformed crews.
FORD ARCHIVES PHOTO

Meanwhile, Stout, with Stout Air Services, had launched his Detroit-Grand Rapids service, and was quietly innovating and setting standards for air passenger travel of the future. As he set out to do, he ran regularly scheduled flights, and within a few weeks had a smoothly running operation. Passengers were still apprehensive of air travel, and on many Stout Air Services flights the planes took off half empty. But Stout stuck to his schedule, and in the first six months of operations there had been no interruptions to service.

When the *new* Ford Tri-Motors became available. Stout purchased these planes to replace the single-engined 2-AT's he had been flying. He bought in all six 4-ATs and, with them, he extended service to Cleveland and Chicago. Stout Air Services made the Cleveland run of 128 miles in 1-1/2 hours. A one-way fare was $ 18; a round trip was $35.

Under Stout's astute management, Stout Air Services pioneered many "firsts" in passenger air service. It was the first airline to serve meals (hot

coffee and sandwiches) to passengers. It had the first "flight escorts." These were attendants who flew every flight, making passengers comfortable, explaining how the airplane operated, and pointing out scenic landmarks en route. It also had the first uniformed crew. This was Henry Ford's idea. "Uniforms give dignity to their profession," he said recommending that both the pilots and the flight escorts be dressed in snappy blue coats and pants trimmed in gold braid.

Stout Air Services also established the first hotel airline ticket office. This began when a hotel porter at Detroit's Book-Cadillac Hotel was persuaded to sell tickets as a sideline. Soon the hotel itself took over, and from the Book-Cadillac Hotel the first airline limousine service also operated. Stout established it to bring passengers from Detroit to Ford Airport.

In addition, Stout Air Services pioneered many operational innovations. It was the first American airline to conduct regular maintenance checks and motor overhauls. It was also the first to set up systematic reporting. Flight reports, maintenance reports, engine-performance reports—all this paperwork, now routine, was introduced by Stout Air Services.

The focal point of all aviation activities at the Ford Motor Company was Ford Airport, whose facilities were used by not only Stout Air Services and Ford Air Transport Service but also by all kinds of visiting aircraft. Henry Ford's invitation to any and all pilots still continued, and in 1926 even the dirigible *Los Angeles* moored at the airport for a few days.

Many improvements in airport facilities had been made. Erected at Ford Airport soon after the new factory had been completed, for example, had been a terminal building. Two stories high and constructed of white brick and stone with a Spanish tile roof, it was the nation's first modern airport terminal.

Both Ford Air Transport Service and Stout Air Services were headquartered here, and each had its executive offices on the terminal's second floor. The first floor was designed principally for passengers. It had a waiting room complete with ticket window, water fountain, and snack bar. For waiting passengers there were upholstered wicker chairs, a fireplace, and an esplanade from which planes could be seen landing and taking off. Later, to accommodate passengers who did not want to travel into Detroit, the Fords built Dearborn Inn across the street. It was the first major airport hotel in the United States.

Ford Airport came alive every morning at 8:30, and the routine seldom varied. The landing field coordinator would walk out toward the runways and look at his watch. "The Chicago mail plane is due to take off in five minutes," he would say.

The Ford Air Transport Service plane would already be at the landing platform, and nearby checking it would be the pilot in his blue-and-gold uniform. A Model T truck would soon chug over to the landing platform, and the driver would step out and unload the mail sacks from the truck, placing them in the Ford Tri-Motor's baggage compartment. The pilot would then get into the plane and the engines would be started. Once warmed up, the plane would taxi down the runway, climbing quickly into the air, heading westward for Chicago.

The landing field coordinator would look at his watch again, and point to the sky. "In two minutes the Stout Air Services plane from Grand Rapids will arrive," he would say. A speck in the sky over the main Ford administration building would grow and become another Ford Tri-Motor. Then, the hum of the plane's engines would be heard, and soon the big airship with corrugated metal sides would come swinging down, land with a slight bounce, and taxi over to the unloading platform. A gangplank would be laid down, and the passengers would climb out.

No sooner would the passengers pick up their luggage and walk into the terminal than the airline limousine from Detroit would head up the driveway to the terminal. Climbing out, passengers would take their luggage into the waiting room, weigh it, and then head out toward the field where the Grand Rapids plane, its engines idling, would be waiting.

Stepping inside, they would take their seats by the half-moon windows and wait for the pilot to step into the plane and go up front to the pilots' compartment. The flight attendant would walk down the aisle, handing out small plugs of cotton. Then, gunning the engines, the pilot would taxi the plane down the runway and take off for Grand Rapids.

Such efficiency of operations depended on many airport facilities that passengers never saw. During the first year of operations at Ford Airport, a complete United States Weather Bureau station—another aviation first—had been installed at the top of the airport's main hangar. Later, for two more firsts, a radio shack and a traffic control system were installed.

Sometimes the improvement in Ford Airport's facilities stemmed from a personal reaction of Henry Ford. Eating lunch one day in the little white house, he noted in the *Detroit News* an article about a recent model-airplane meet held at Ford Airport. A national contest, with two entries from each state, the event had been backed by Edsel Ford, who not only donated the facilities of Ford Airport for the meet, but also tendered a banquet to the contestants in one of the hangars. During the meet it had unfortunately rained heavily, and the *Detroit News* reporter covering the contest had devoted almost as much space to the condition of the landing field as to the details of the meet. "The model-airplane meet was held in a bath of mud and soggy areas of the sunken airport," he had written.

Ford bristled. Leaving his lunch half eaten, he strode into Mayo's office. "Cover the field with concrete," he ordered. The next month Ford Airport had the world's first concrete runways.

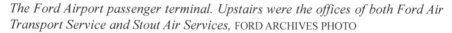

The Ford Airport passenger terminal. Upstairs were the offices of both Ford Air Transport Service and Stout Air Services, FORD ARCHIVES PHOTO

An interior shot of the Ford Airport passenger terminal. It featured wicker chairs, a water fountain, and a ticket window. FORD ARCHIVES PHOTO

Another innovation was aviation's first successful radio-beam guidance system for planes. Both the army and the Bureau of Standards had previously conducted experiments on such systems, but nothing practical had resulted. Two Ford engineers, C. W. Thomas and Eugene Donovan, picked up the problem in 1926. Within months they had set up a workable guidance system at Ford Airport. Planes fitted with the proper equipment were able to beeline in by radio beam from as far away as fifty miles. Then, in February, 1927, a Ford Tri-Motor, guided solely by radio beam, flew from Ford Airport to Wright Field, Dayton. Adaptations of the Ford system were later taken up by all United States airlines, and although the Ford Motor Company held the basic patents, no royalties were ever asked of either those who manufactured the systems or those who used them. It was another free contribution of the Fords to aviation.

The pilot of the Ford Tri-Motor making the radio-guided flight from Ford Airport to Dayton was Harry Brooks, who was rapidly becoming an important figure in aviation at Ford. Brooks was the son of a longtime personal friend of Henry Ford. The senior Brooks was a fiddler, and in Ford's view, the best fiddler in the world. Twelve months previously, Brooks had come to Ford and said he could do nothing with his boy. "All he ever thinks about is airplanes and flying," the father said.

Ford understood the problem. Evidently Brooks equated his son's interest in aviation with being a ne'er-do-well. "Send him to me," Ford said. "I'll see what I can do."

When Harry Brooks appeared, Ford gave him a job in the Airplane Manufacturing Division. The youthful Brooks turned out to be a remarkably good pilot. Within a short time he was the company's chief pilot.

Despite the subsidies given by the federal government to the airlines for carrying mail, despite the millions of dollars being poured into aviation by Henry Ford, commercial aviation in the United States in 1927 still lagged. Most airlines were still flying ancient, World War I-vintage planes. Most airports were little better than cleared farmers' fields. Most of the public had no interest in flying. If anything, Anthony Fokker's analysis of the state of commercial aviation—as being not far enough along to support large-scale plane production—was accurate.

If United States commercial aviation was ever to develop into a full-fledged industry, something dramatic had to take place. And then on May 21, 1927, that "something" did take place. That is when the wheels of a silver-colored Ryan monoplane named the *Spirit of St. Louis* touched ground at Bourget Field outside Paris, and a tired Charles A. Lindbergh called out, "Any mechanics here?"

While flying airmail for Robertson Aircraft, Lindbergh had first heard of the $25,000 prize being offered by Raymond B. Orteig for the first nonstop flight from New York to Paris. With the backing of friends Lindbergh bought the Ryan monoplane, and even before he left for Paris he set a transcontinental record flying from San Diego to New York.

A week later, on May 20, 1927, at 7:52 A.M., he took off for Paris, landing the following day at 10:21 P.M., Paris time.

A huge crowd had assembled at Bourget Field to await Lindbergh's arrival. The minute he landed that evening, thousands surged forth. Lindbergh felt the Ryan's fuselage crack as the crowd pressed forward and began ripping fabric off the plane for souvenirs. Someone grabbed Lindbergh's legs, and he was dragged out and raised aloft over the screaming crowd. Two French pilots managed to rescue the American aviator and hurry him to an empty hangar. Putting out the lights so the crowd would think he was not there, they waited in silence until everyone dispersed. Then they drove Lindbergh over back roads to the American Embassy.

Lindy's great feat electrified the nation. Congress raised him to the rank of colonel and awarded him both the Congressional Medal of Honor and the Distinguished Flying Cross. Tremendous receptions awaited him in New York City and Washington, D.C., and for months the excitement continued. A national hero overnight, Lindbergh capitalized on his sudden fame to promote aviation. Once the parties were over he took the *Spirit of St. Louis* on a triumphant nationwide tour to seventy-eight cities and every state in the Union.

Everywhere Lindbergh traveled, people went wild. The bankers, industrialists, businessmen, and the public—all those who just a few months before had ignored aviation—now hurried to jump on the aviation bandwagon. For months the nation's newspapers filled their front pages with news about planes and pilots. Three months after Lindbergh's epic flight, *The New York Times* was still running dozens upon dozens of articles daily about aviation. In its August 27, 1927, issue, for example, there were no fewer than seventy-four different articles and features on aviation and flying, among them: a report that the airplane *Pride of Detroit* was about to hop off in a race around the world; a story about the memorial rites held at sea for three fliers lost in the Dole Pineapple Company's Pacific Air Derby from the United States to Hawaii; a feature on Paul Redfern, an aviator lost on a flight from New York to Brazil; and a story of two airmail pilots about to fly from the United States to Rome in a Fokker monoplane named *Old Glory,*

On and on the aviation items went in that day's *New York Times:* two pilots killed at a Teterboro, New Jersey, air circus when they tried to change parachutes in midair; transatlantic flyer Clarence Chamberlin honored by Iowa College; Lieutenant A. J. Williams delayed in testing a new biplane; Umberto Nobile planning a dirigible flight over the North Pole; an army pilot arrested for flying too low over the Capitol in Washington, D.C.; New York City seeking land for a new airport; a family of eleven taking an air ride; General Electric introducing a new radio transmitter for planes; a Bolivian airline reporting successful operations; two flyers leaving Curtiss Field, New York, for Windsor, Ontario; two Royal Air Force officers postponing their flight from London to Canada; South American nations planning transatlantic mail service; and Lindbergh greeted in Fargo, North Dakota.

Aviation at long last had arrived, and the Fords—with their production lines all set up to manufacture Ford Tri-Motors—were ready.

Despite their efforts in popularizing aviation, neither Henry Ford nor Edsel had ever been up in an airplane. This state of affairs changed on August 11, 1927, when Lindbergh visited Ford Airport as part of his nationwide tour. Everywhere Lindbergh had flown in the *Spirit of St. Louis* he had drawn tremendous crowds, and Ford Airport was no exception. When his small silver-colored Ryan monoplane appeared in the sky over Dearborn, thousands were on hand to greet him. The welcoming committee was headed by the two Fords and by Mayo and Stout, all dressed in white tropical suits and Panama hats.

When Lindbergh climbed out of the *Spirit of St. Louis* he looked just like his photograph—tall, slim, lanky, with that famous leather flying cap on his head, complete with ear flaps and goggles.

After pleasantries were exchanged, Lindbergh on the spur of the moment asked Henry Ford if he would like to go up in the *Spirit of St. Louis.* To everyone's surprise, Ford, who previously had resisted all opportunities to fly, said, "Yes."

Lindbergh's plane had never carried any passengers before, so Ford mechanics had to install an extra seat. But Ford was game. Climbing in, he settled down, and Lindbergh followed, easing himself into the pilot's seat after first making sure Ford was secure. Lindbergh then started the motor, and the *Spirit of St. Louis* was soon in the air. For fifteen minutes Lindbergh stayed aloft, climbing to about two thousand feet. Then he returned.

"A great trip," Henry Ford exclaimed. "I certainly enjoyed it."

Lindbergh then took Edsel up for his first plane ride, and Edsel was equally enthusiastic. After the flight Lindbergh was persuaded to fly a Ford Tri-Motor. All shined up and ready, the plane had been sitting on the other runway. Henry Ford climbed aboard for this flight too. Accompanying him were Mayo, Stout, several newspapermen, and Major Thomas Lanphier, commandant of the army's nearby Selfridge Field.

Lindbergh visits Ford Airport, and Henry Ford takes his second plane ride. Boarding the 4-AT are, left to right, William B. Mayo, Henry Ford (entering the plane), Lindbergh, Stout, and Edsel Ford. FORD ARCHIVES PHOTO

With Lindbergh as pilot and Harry Brooks, Ford's chief pilot, as copilot, the big corrugated-metal plane was soon airborne. Turning over the controls to Brooks, Lindbergh walked back into the passenger cabin.

"How high are we?" Ford asked him.

"About three thousand feet," Lindbergh said.

"Gosh, it's a long way down."

For nearly an hour the Ford Tri-Motor cruised over Detroit, circling Belle Isle Park and various Ford plants, including River Rouge.

"I wouldn't mind taking a spin every day," Henry Ford said when they returned. "It's like going on a picnic."

But he never flew again.

Ironically, the excitement over Lindbergh caused the public to lose interest in other feats of aviation heroism. After Lindy's historic New York to Paris flight, dozens of other aviators also tried to fly across the Atlantic. As many as ten were successful, but they were quickly forgotten. As aviation historian Martin Caidin explained: "The time for such spectacular flights has passed. . . . They proved nothing and nobody was interested." So uninterested was the public now that the editor of a leading aviation magazine announced he would no longer publish information on such flights. In 1929, when the *Yellow Bird* made a successful flight from New York to Paris, more than a thousand gendarmes were at Bourget Field when the plane landed—but only three hundred spectators. "All the first flights below the stratosphere had been made," Martin Caidin said. "There were no more big money prizes to tempt the foolhardy. The pioneering . . . had been completed. There was no more room for men who flew by the seat of their pants."

The decline of the individual aviation hero was matched by increased interest in commercial aviation. Suddenly the public was eager to invest in aviation stocks and to urge their communities to build municipal airports. Suddenly, too, people were no longer afraid to fly.

Chapter Seven
Commercial Aviation Takes Off

W ith the climate more favorable than it had ever been before, the airlines began to expand their routes and services. Stimulating their growth was a decision of the United States Post Office to turn over to commercial airlines the one remaining major route still being flown by the Air Mail Service—the Columbia Route, between New York and San Francisco.

Rather than let one airline fly the entire route, the Post Office decided to divide the route in half. One airline would be awarded the mail contract from New York to Chicago; another, the contract from Chicago to San Francisco. The company that won the eastern half of the route was National Air Transport. The western half went to Boeing Air Transport.

Boeing, fast becoming a major factor in commercial aviation, was actually one of the nation's oldest aviation companies. It traced its beginnings to 1915 when William E. Boeing, son of a wealthy West Coast lumberman, purchased a Martin pontoon plane to use in flying to the British Columbia lakes and rivers where he liked to fish and hunt. Soon after he purchased the plane, it cracked up. When Boeing ordered another plane, he was told a replacement would take six months to deliver. "God, we can build a better airplane than this and do it faster, too," Boeing said to a flying friend, a retired navy commander named Westervelt. The two built a flying boat, and so well did it fly that they decided to enter the airplane-manufacturing business. Acquiring an old boatbuilding factory on the Duwamish River near Seattle, they set up a shop. Later, with a World War I aviator named Edward Hubbard, they began flying international mail between Seattle and Victoria, British Columbia. This was in 1919, six years before the Kelly Act, which concerned domestic mail only. Thus, their service was unsubsidized, but even so, they managed to make a profit.

Preceding Page: The Boeing 40-B Mailwing. An improved verson of the 40-A, it carried mail, freight, and, in small cabin compartment of the pilot's cockpit, several passengers. UNITED AIRLINES

David A. Weiss

William E. Boeing (right) and his 1916 plane. Standing beside him is his first partner, Edward Hubbard. THE BOEING COMPANY

Boeing saw an opportunity in the Columbia Route to expand his airline, and once he got the contract, he moved into action. Copying the move C. M. Keys had made in organizing NAT, Boeing hired as head of the new operation the man who headed the Air Mail Service—D. B. Colyer, the second assistant postmaster general. Colyer knew personally the pilots who had flown the route for the Post Office. After he went to work for Boeing, he turned around and hired them for the airline. Next, in another strategic move, Boeing established his headquarters for the new service in Salt Lake City. Western Air Express already flew here, and Varney Air Lines was about to. Boeing calculated (rightly, it turned out) that he could link up with both airlines, and thus further extend his own service.

Boeing started operations on his Columbia Route segment in July, 1927. For planes, he used aircraft of his own manufacture. These were Boeing 40-A mail planes. Pilots rode up front in open cockpits. Passengers were squeezed into a tiny cabin between the wings. No more than two could be carried on each flight.

Other airlines were also expanding. Varney, which soon connected to Boeing and Western Air Express in Salt Lake City, got another feeder airmail contract between Spokane and Seattle. By the end of 1928, this airline was earning almost a million dollars yearly from its airmail contracts.

Another new airline had begun making a name for itself. This was Pan American Airways, soon to become the biggest giant of all. Pan Am got its start in 1927 when the airmail contract from Key West, Florida, to Havana, Cuba, came up for grabs. Three companies bid for it. One, Florida Airways, was established, but this company had come on hard financial times, and was heading toward bankruptcy. Another was Pan American Airways, recently established by a group headed by an ex-army pilot, Captain John K. Montgomery. A third company was led by Juan Trippe, son of a Manhattan banker and a former Yale football player. Descended from a seventeenth-century Maryland seafaring family, Trippe had been a naval pilot in World War I and had been a flying enthusiast ever since. With several Yale friends Trippe had organized Long Island Airways, but had difficulty expanding this airline. He tried to become affiliated with Colonial Airways, which operated between New York and Boston. When these negotiations failed, he organized a new company—Aviation Corporation of America—for the purpose of buying into other airlines. Backed by John Hay Whitney, W. Averell Harriman, and William H. Vanderbilt, the new company had a capitalization of $300,000. With this financing, Trippe was working out an arrangement to take over Pan American Airways when Pan Am received the Key West-Havana mail contract. In October, 1927, Pan Am, now owned and operated by Trippe's group, started service in a Fokker trimotor.

All these airlines, with the exception of Boeing, which used its own planes, began to purchase Ford Tri-Motors. They were the answer to most of the problems then in passenger and freight transport. As Martin Caidin said: "Americans, mostly businessmen, wanted to buy tickets to ride in an airplane. After all, if Lindy could fly all the way to Paris behind one engine, it must be safe to fly from Detroit to Chicago in a plane with three engines—particularly in a plane made by Ford."

By now, Ford and Stout had more than demonstrated that commercial aviation was feasible. Stout Air Services had proved that passenger travel by air was safe and comfortable. Ford Air Transport Service proved that transporting freight by air was practical.

Ford never lost an opportunity to publicize the Ford Air Transport Service, which was still the world's largest private airline. True to his word when he established his service, he issued periodic reports on its operations and facilities. In mid-1927, for example, he announced that the airline had now been extended to Buffalo, and that it had flown safely more than half a million miles and carried more than three million pounds of freight. During this time, the company had made 1,492 scheduled flights, of which 1,467, or 98.3 percent, had been completed. Chief causes of non-completion of flights were engine trouble and bad weather. As for the cost of flying freight, this averaged out to 13.8 per pound.

The chief purpose of Ford Air Transport Service was to transport auto parts to the various Ford assembly plants in the midwest. Almost every component that went into a Model T had been flown: wire wheels, cushions, carburetors, and tires. Manufacturing equipment, such as jigs, gauges, wrenches, screws, and bolts, had also been freighted by air. Once, a thousand-pound lathe had been flown from Detroit to Cleveland.

But Ford used its private airline in other ways too. Pilots were trained on the line, as were field service mechanics. Also, in the course of operations, valuable experience was gained in navigation, radio, meteorology, and in the art of handling heavily loaded transports in all kinds of weather. Ford Air Transport Service also served as a testing ground for proposed modifications in the 4-AT model. Variations in design, new engines, improved controls—experiments were continually being tried along the airline's routes.

In July, 1927, Edsel paid a visit to President Calvin Coolidge, who was vacationing in the Adirondacks. The purpose was to report to Coolidge on the status of American aviation and, in particular, the progress made by Ford. "I believe the future of flying machines lies with multiple-engined ships made of metal," Edsel told Coolidge. "These planes make flying safer and they also give a longer life to the plane."

Edsel said that the cost of the Ford Tri-Motors would soon be reduced to $42,000 per plane, and that after the first hundred were produced, a further reduction of 25 percent would be made. "These planes will soon be so inexpensive that air routes will be able to operate at a profit," he said. Edsel stressed, however, that for the time being it seemed unlikely that any airline could be put on a paying basis without airmail contracts. "Hopefully this situation will change," he said.

Soon after, the third Reliability Tour was held. Now called National Air Tour, it flew a longer distance (four thousand miles) than in previous years and visited more cities. Again Edsel donated the trophy, and both he and his father were at the finish line when the planes, bucking a windstorm, flew in to Ford Airport from Grand Rapids on the last day. Again the Ford entry failed to win. This honor went to Eddie Stinson, flying one of his own Stinson Detroiters.

Soon after, the Fords were visited by Commander Richard E. Byrd, the nation's most popular aviation hero next to Lindbergh. Byrd too had competed for Orteig's New York-to-Paris nonstop prize, but, because of a series of mishaps, had been unable to take off for Paris until a month after Lindbergh had. Unlike Lindbergh, who had flown solo, Byrd took along three passengers. He crossed over successfully in his plane, a Fokker trimotor he named *America,* but upon reaching the French mainland ran into heavy fog and rain and was unable to land. Circling blindly for hours, Byrd flew another five hundred miles before running out of fuel. He was finally forced down near the coast. Landing was dangerous, and the *America* was wrecked, but no one was injured.

Now Byrd had a new project in mind. The first man to fly over the North Pole, he also wanted to be the first to fly over the South Pole. As before, he had come to the Fords for financial assistance.

Preparing to take off is the America, *the Fokker trimotor in which Byrd attempted a nonstop flight over the Atlantic Ocean with three passengers.* PICTURE COLLECTION, NEW YORK PUBLIC LIBRARY

At the Ford home Byrd and Henry Ford traded stories. The elder Ford talked about his first automobile and showed it to the explorer ("my thrill for the day," Byrd said). Byrd told Ford about his recent flight across the Atlantic in the *America.* The conversation continued so long that Byrd had no time to fly one of the Ford Tri-Motors, which he had planned to do. But he was able to see the press, and at this time Edsel announced that the Fords would give Byrd a 4-AT for his polar flight and back him in every other way they could. "We are doing this because he is a gentleman," Edsel said, "and because what he is doing is a great service for aviation and the world."

Lindbergh was also making news. No sooner did he complete his sensational tour of the United States than he started an even bigger goodwill tour south of the border. President Calles of Mexico had invited him to visit Mexico City, and Lindbergh decided to fly from the States to the Caribbean, Central America, and parts of South America. Before he returned to the United States, he would fly almost eight thousand miles and visit sixteen countries.

On December 13, 1927, Lindbergh took off for Mexico City. He made his headquarters there at the American Embassy. The United States ambassador to Mexico was Dwight Morrow; his daughter Anne would marry Lindbergh eighteen months later.

A few days after Lindbergh arrived in Mexico, a news story appeared in the Detroit newspaper. Mrs. Evangeline Lindbergh, Lindy's mother, had declined an invitation to travel to Mexico City to be with her son during the Christmas holidays. A chemistry teacher in a Detroit high school, Mrs. Lindbergh said that if she traveled to Mexico by the usual railroad route, she would be unable to get back in time for her classes when they resumed after the New Year.

Stout, reading about this in the newspaper, got an idea. Why not offer to fly Mrs. Lindbergh back and forth to Mexico City in a Ford Tri-Motor? Approaching Henry Ford, Stout was told that the mountains encountered en route to Mexico City would probably make such a flight impossible, and, besides, Mrs. Lindbergh would not want to fly. "But you check with Edsel," Henry Ford said.

Stout did this, and was told the same thing. "I think the mountains would be too much," Edsel said. "But you might ask Lejous, the head of our Mexican office, who is now in Dearborn."

Locating Lejous, Stout found out just the opposite. Lejous had flown up in a Ryan monoplane. He enjoyed the flight and it was not dangerous at all. In fact, he would like very much to fly back to Mexico City in the Ford Tri-Motor.

Telephoning Mrs. Lindbergh, Stout found she could not make up her mind. She suggested Stout contact her son in Mexico City and ask him. This Stout did by telephone, and Lindbergh hesitated. Then Stout said that Harry Brooks himself would fly the plane. Lindbergh gave his approval. "If Brooks is the pilot, I'll agree to it," he said.

David A. Weiss

Checking the sun-compass reading while sitting on top of the America, *Roosevelt Field, New York, May 16, 1927, (left to right) Anthony Fokker, Bernt Balchen, and Richard E. Byrd.* UNITED PRESS INTERNATIONAL PHOTO

The Ford Tri-Motor was scheduled to take off from Ford Airport on December 18, 1927. It was a cold windy day. In deference to Mrs. Lindbergh, the plane would be boarded inside the hangar. Mrs. Lindbergh had been brought to Ford Airport in a limousine. A tall elderly woman, she wore a black, fur-trimmed coat to which was pinned a white gardenia. On her head was a small black felt cap, and on her feet, galoshes. Greeting her were Mr. and Mrs. Stout, who would make the trip with her, Henry Ford, who had come to see her off, and the parents of Harry Brooks.

The party posed for newspaper photographs, and then walked over to the hangar. Brooks had already been warming up the Ford Tri-Motor's engines. The plane had been specially fitted for the trip. Its passenger cabin was electrically heated, and some of the wicker chairs had been removed so sleeping berths could be installed. Ford helped Mrs. Lindbergh inside, and once everyone was secure, Brooks taxied the plane outside the hangar toward the runway, where the big 4-AT soon took off.

The plan was to fly to Mexico City in a series of hops. Saint Louis would be the first stop, then San Antonio, then Tampico, and finally Mexico City. The Ford Tri-Motor made Saint Louis right on schedule, and then headed south. By nightfall, it had reached San Antonio, where everyone got out and drove into the city to spend the night. At the hotel Mrs. Lindbergh was given the presidential suite, and the Stouts the bridal suite. After everyone had retired, the Stouts heard a knock on their door. It was Mrs. Lindbergh. "It's too lonesome over there, and the many flowers make it look like a funeral," she said. "Can I spend the night in your suite?"

The next day they arrived in Mexico City. Lindbergh was the idol of the capitol. "The Ambassador of the Air," the press described him, and Mexican newspapers ran page after page telling of his aerial feats and how he had taken up President Calles and ex-President Obregon on their first plane flights.

Three days after Christmas, Lindbergh took off for Guatemala, and Mrs. Lindbergh and the Stouts flew back to Detroit in the Ford Tri-Motor. The trip was uneventful until the plane reached Springfield, Missouri, where it ran into snow. Brooks decided to land in Chicago, but with night closing in, he could not find the airport. Circling low, he kept flying, trying to spot some electric signs that would give him some bearings. Suddenly seeing a white space, he came down for a landing. Sliding in, he landed the plane successfully, only to find his white space was the Robey Race Track. Both the plane and its passengers were safe, but later they learned that another plane had crashed there several hours before.

Aviation was now beginning to boom. Although aviation securities on Wall Street accounted for only 0.2 percent of all new issues in March, 1928, by October, 1928—less than a year and a half after Lindbergh's New York to Paris flight—they accounted for 2.8 percent of all new issues, and by March, 1929, for 11.2 percent. During the same period, the aviation stock index rose from 96.4 to 1,260—an increase of more than a thousand points.

The number of airlines being organized also increased. In 1927, twenty-three new airlines were established. Two years later, ninety-seven airlines were formed. More and more, too, big business was becoming involved in aviation. Citing the Fords as an example, the DuPonts announced their entrance into the field. Not only were they building an airport in New Castle, Delaware, they were also investing a million dollars in a new aircraft-manufacturing company to be headed by Giuseppe Bellanca.

The growing interest in aviation produced more sales for Ford Tri-Motors. In 1927, the Airplane Manufacturing Division sold twelve 4-ATs. The following year, it sold thirty-nine 4-ATs—more than three times the previous year's sales. To keep pace with this demand, Ford announced it was expanding its production facilities. It would soon, the company said, be able to produce one Ford Tri-Motor a day.

As in previous years, some trimotors were bought by corporations to serve as executive airplanes. The Nekoosa-Edwards Paper Company, for example, purchased one, as also did Wright Aeronautical Company and Union Electric Light and Power Company. Another company buying a Ford Tri-Motor was Reid Murdock and Company, manufacturers of Monarch Foods. Their 4-AT was stripped down and lined with shelves, which were filled with Monarch canned goods. Called "The Flying Grocery Store," the plane flew around the midwest for years, promoting Monarch foods.

Many of the new smaller airlines being organized also purchased Ford Tri-Motors, These included Gray Goose Air Lines, Chicago; Rapid Air Lines, Rapid City, South Dakota; Jefferson Airways, Minneapolis; Queen City Airlines, Cincinnati; Universal Flyers, Washington, D.C.; and Sunbeam Air Transport, Denver.

Another airline purchasing a Ford Tri-Motor was Scenic Airways, later Grand Canyon Airlines. Headquartered at Grand Canyon, this airline used its plane in many different ways. It toted in supplies to isolated ranchers. It conducted aerial reconnaissance along the big gorge. It later shuttled passengers between Canyon City and Hoover Dam. Mostly the Ford Tri-Motor was used to fly tourists on a one-hundred-sixty-mile run through Grand Canyon. On the trip, pilots routinely would drop down into the rock cut three thousand feet below the canyon rim. "We scared the daylights out of the passengers," said Ed Campbell, a former Scenic Airways pilot. "We'd drop through the clouds and fly down a tunnel between the river and the canyon walls. It was spooky, but safe enough. The Ford was a very forgiving airplane. She could get you out of almost anything you could get her into."

Another sight-seeing line using a Ford Tri-Motor was Sky View Lines, Niagara Falls. For five dollars a head, tourists were flown over the falls. One Sky View pilot was Harold E. Gray, later a president of Pan Am. Originally Gray had worked as an aeronautical engineer for the Stinson airplane factory. Talked into leaving there to become an instructor in the Ford Pilot-Training School, Gray let himself be lured away again by Sky View Lines when it bought its Ford Tri-Motor. The inducement was a four-hundred-dollar monthly salary, high for those days.

Most Ford Tri-Motors were purchased by the larger airlines. In the beginning these companies had had motley assortments of planes in their fleets. Robertson Aircraft in 1927 was flying five De Havilland DH-4s, three Boeing Mailwings, one Standard, one Waco, and one Travel Air.

But gradually the major airlines turned to planes of one manufacturer and one type, and, more often than not, Ford Tri-Motors were selected. Thus, other airlines besides the ones directly involved with Ford (like Stout Air Services, Maddux Air Lines, and Ford Air Transport Service), began purchasing Ford Tri-Motors in quantities of more than just one or two.

One of these airlines was Pan American Airways, which was undergoing tremendous expansion. From the company's one small airmail route between Key West and Havana, it was now flying all over the Caribbean as well as to Central America and parts of South America. Juan Trippe had signed on Lindbergh as technical adviser, and Lindbergh recommended that, for rapid growth, the airline wherever possible establish its landing points on water, thus avoiding the delays and expense in building airfields. As a result, most of Pan Am's planes were Sikorsky flying boats, but as the airline's routes continued to expand into Nassau, Cuba, Mexico, Florida, and the Canal Zone, it started to purchase land planes as well, and these were mostly Ford Tri-Motors.

About the only major United States airline not purchasing Ford Tri-Motors was Boeing Air Transport, which used planes of its own manufacture. Boeing was expanding almost as rapidly as Pan American, Soon after it started flying mail over the western half of the Columbia Route, Boeing acquired Pacific Air Transport, which gave it a Los Angeles-Seattle route, and then it began negotiating for other airlines.

Key to the Boeing growth was the good profits it made on mail contracts, and key to these was the use of a new engine on its Boeing Mailwing planes. Just as the Wright Whirlwind engine had been an improvement over the Liberty, so was a newly developed Pratt & Whitney Wasp engine an improvement over the Wright Whirlwind.

Pratt & Whitney was a newcomer to the field of aviation, and Wright Aeronautical Corporation had itself to blame for the competition. The man chiefly responsible for developing its Whirlwind engine was Frederick D. Rentschler, onetime head of Wright Aeronautical Corporation. After the Whirlwind had been introduced, Rentschler was approached by the company's chief engineer. He had an idea for an improved engine. But when Rentschler tried to get approval from Wright's board of directors to develop this new engine, the board turned him down. Why should Wright spend money and effort to develop a new engine when it already had the best engine in the industry? Rentschler's response was to leave Wright and organize a company of his own to develop the new engine. Setting up a factory in a tool plant abandoned by a defunct company called Pratt & Whitney, Rentschler took over the company's name as well as its factory.

Within a short time, the Pratt & Whitney Wasp engine was ready, but the navy tied up production, ordering Wasp engines for all its carrier planes. Learning of the engine, Bill Boeing persuaded Rentschler to sell him twenty-five engines, navy contract or not. Around these new Pratt & Whitney engines Boeing redesigned his Boeing 40-A Mailwings, increasing their payload from a thousand pounds to fifteen hundred pounds and enabling them to carry more passengers.

Ford, meanwhile, was busy turning out 4-ATs. Although the planes were made on a production-line basis, many minor variations and modifications were made. Through the months, more streamlining was added, controls were improved, and a new tail wheel was introduced. Changes were also made in the passenger cabin. The half-moon windows were replaced by rectangular windows, more soundproofing was added, and the chairs were changed from wicker to upholstered leather.

In many cases, special interiors were designed for special needs. In the case of one 4-AT destined for use by Ford Air Transport Service, the interior was completely stripped so the plane could carry more freight. Other times, particularly when the plane was purchased by a corporation as an executive plane or by an individual, the interior was fancied up, and the plane was equipped with dining table, glassware, and china.

With production and sales of Ford Tri-Motors at satisfactory levels, Henry Ford continued to explore other aviation opportunities. Returning once more to his interest in dirigibles, he resumed negotiations with the navy and finally persuaded it to give Aircraft Development Corporation a contract to build a small all-metal dirigible. He also began to think that Ford Motor Company should build other kinds of airplanes besides the 4-AT Ford Tri-Motor.

Ford's thoughts went toward two extremes. On one hand, he was thinking about a huge one-hundred-passenger airplane. On the other hand, he had the idea of building a "fliwer" plane for just one or two people or possibly a family.

His thoughts about a giant one-hundred-passenger plane stemmed from his conviction that commercial aviation without mail contracts—relying on passenger income alone—would never be economically feasible until at least one hundred passengers could be carried on each flight. When the first 4-AT had been successfully tested in June, 1926, everyone had walked away from the landing field, smiling and congratulating one another. Ford had been pleased too, but when Stout approached and asked, "What do you think of it?" Ford answered, "Not big enough."

"How big should it be?" Stout asked.

"One hundred passengers."

Stout smiled. He would agree that such a giant plane would be a boon to aviation. But such a giant land airplane could not be built within the framework of existing technologies and the engines then available.

"I am sure we will be able to get a hundred-passenger airplane eventually," Stout said. "From our eleven-passenger 4-AT we will be able to build a thirty-passenger plane and, from this, go on to a fifty-passenger plane, and then, before we know it, we will be able to build a hundred-passenger plane."

"The gradual approach is not for me." Henry Ford shook his head. "Rather than keep building small planes from which we would learn nothing, I would rather build a big plane, even if it did not fly."

Although Stout could not visualize such a giant plane at that time, Mayo did, and, soon after the 4-AT was introduced, Hicks was given instructions to start designing a one-hundred-passenger plane. Secrecy veiled the project, but reports filtered out from time to time. Mayo, in an expansive mood after he landed in San Diego with Jack Maddux's initial 4-AT Ford Tri-Motor, leaked the first public announcement of the plane to a local newspaperman. "Yes, we are building a giant 100-passenger plane," he said. "It will have six air-cooled engines, each of which will deliver 1,000 hp, and it will fly 125 miles per hour and have a range of 700 miles."

Mayo said the Fords had already spent more than a year designing the giant plane, and that construction of it would be completed within the next year. "This plane will be the Pullman car of the air," he said. "It will have sleeping quarters for all passengers, and all the other facilities and comforts of the finest rail coach."

According to Mayo, the probable cost of the plane would be $100,000, and it would eventually be mass-produced. But twelve months passed, and nothing more was announced about the plane, although rumors persisted that the Fords were still working on it.

Meanwhile, Ford Motor Company was also working on Henry Ford's other pet project—a "flivver of the air." Ford had justifiably never forgotten the impact he had made on the American public with his Model T. His ambition, as reported by many of his friends and associates, was for the Ford Motor Company to build—under Edsel's direction—a flying machine that would be similarly inexpensive, durable, and mass-produced. He saw the plane as being used by families and individuals in transporting themselves from city to country and from home to work.

The Ford "Flivver of the Air." Designed for millions, it was flown by only two people—Lindbergh (shown here) and Harry Brooks. FORD ARCHIVES PHOTO

So, while the Airplane Manufacturing Division was producing Ford Tri-Motors, it was designing on paper dozens of experimental flying flivvers of all sizes, shapes, and speeds. For a time the company concentrated on a "sedan plane" or "family plane" that would carry from two to four passengers. The idea had been to scale down the 2-AT *Maiden Dearborn* type of plane to a smaller version that would be used by "plain people." At one point Edsel gave out some information on this plane, saying that the company was definitely about to produce it, and that it would sell for as low as $2,000 to $3,000 when manufacturing got onto a production-line basis. "All we need is a good air-cooled 200 hp engine," he said, "and this we are now working on, and should soon be able to perfect."

But nothing more was heard of this plane, either.

Then the company concentrated on another type of flivver plane—a one-passenger plane. This was Henry Ford's own idea. He saw a need for a tiny craft that an individual could use for his own transportation. In visiting Stout's office one day, Ford asked Stout what he thought of such an idea.

Stout's answer was to ask a question. "If the plane holds only one, how could anyone be taught to fly it without killing himself?"

Ford never mentioned the tiny plane to Stout again. The next thing Stout knew, Harold Hicks, Shorty Schroeder, and several others in the Aircraft Manufacturing Division were beginning to build such a plane in another part of the Ford plant in Dearborn.

Although information on this tiny plane is sketchy, two seem to have been built, both of which were single-place monoplanes with low wings and two-cylinder engines. On July 30, 1926, Henry Ford's sixty-third birthday, the press assembled at Dearborn for its annual interview. Reporters asked the usual questions about the Model T ("We have now produced fourteen million of them," Ford said), and whether Ford felt his age ("The first time I was called old was when I was thirty and worked for Detroit Edison"). But the highlight of the interview was when Ford took the newspapermen over to a Ford Airport hangar, pulled a curtain, and unveiled the flivver plane behind it. The tiny monoplane was so small it seemed almost like a toy. It had a wingspread of only twenty-two feet, and its fuselage was only fifteen feet long. No one there had ever seen a smaller single-seated plane.

Ford said the company expected to produce the plane in large quantities using mass-production methods, but that before this could be done, the company had to perfect a new small engine.

Turning to the engine on the plane, Ford said this was a French Anzani air-cooled engine, and that it was only temporarily installed. "Our engine will be just as powerful, but much lighter," he said.

In the months that followed, Hicks did develop a light two-cylinder 36 hp engine, and, after this was installed, the plane was seen flying on experimental low-altitude flights over the Detroit River. On one occasion it even raced a Gar Wood speedboat. Weighing only 370 pounds and capable of flying more than 100 mph, this air flivver was flown by Harry Brooks, who performed spectacular stunts in it and also broke the world's distance record for small planes. Brooks was the only Ford pilot who ever did fly the tiny plane, and only one other person flew it. This was Lindbergh. Asking questions about the small plane the day he visited Ford Airport, Lindbergh was invited to take it up. The tall aviator hero could hardly fit into the plane. To reach the rudder bar without hitting the dashboard with his knees, he had to remove his shoes. The flight was successful, and Lindbergh congratulated those responsible for designing the tiny air flivver.

Experimentation on new planes like the flivver plane could not always be carried out smoothly within the Ford organization. Once when Shorty Schroeder was working on an improvement to the flivver plane's controls, he placed a requisition for a dozen one-cent cotter pins. At the same time the $250,000 dirigible mooring mast at Ford Airport was being improved. Schroeder filled out one follow-up requisition after another, but no cotter pins were forthcoming. He was still waiting when he discovered that the work on the mooring mast had been completed.

Burned, Schroeder could not restrain himself when Henry Ford one day asked him how things were going.

"It's easier to get a mooring mast around here than a few cotter pins," he told Ford.

Ford wagged his finger. "No, it's just as easy to get a mooring mast as a box of pins—*if you're right.*"

Schroeder never did get his cotter pins.

One administrative difficulty was that, through the years, friction had developed between the company's airplane division and its automobile division. It had started soon after Ford had taken over the Stout Metal Airplane Company. Charles Sorensen, a rising young manufacturing executive in the automobile division, had walked over to Ford Airport one day and inveigled Harry Brooks to give him an airplane ride. Brooks was happy to oblige, but Henry Ford, hearing about the incident, telephoned Sorensen and told him to "keep away from the airplane division."

From that time on, despite Hicks's best efforts, the auto division made life difficult for the airplane division. This often hindered Hicks's operations because many design changes had to be cleared by the company's engineering department, which was attached to the auto division. In trying to improve the engine performance of the 4-AT, for example, the airplane division got the idea of installing heaters around the carburetors. But the Ford engineering department turned it down. No heaters were needed for the carburetors in auto engines. Why should they be needed in airplane engines? Hicks tried to explain that airplanes flew at high altitudes where freezing temperatures were often encountered. But the engineering department still withheld its approval.

Hicks was stymied until one day he noticed a Ford engineering-department executive sneaking aboard a new 4-AT about to be tested. The executive obviously figured he was going to get a free plane ride despite Henry Ford's restriction. Hicks got an idea and told it to Shorty Schroeder, who was about to take the plane up. So that he would not be noticed, the executive had shut himself up in the plane's lavatory. Schroeder, taking his seat at the controls, pretended he did not know the executive was aboard. Once the plane was aloft, he told the copilot to "take her up as high as we can go." Then, walking back to the lavatory, he opened the door and told the executive, "We know you're here, so you might as well come out." Higher and higher the plane climbed. The plane's interior began to get bitter cold, and the executive started to shiver. When they got to fifteen thousand feet, the plane's engines started to smoke and sputter. They sounded as if they were about to quit. "You see why we need heaters for the carburetors," Schroeder told the executive.

When the plane landed, the executive went back to his office with Schroeder and filled out the necessary papers. From then on, Ford Tri-Motors had carburetor heaters.

Not long after, the Ford Motor Company was catapulted into a non-aviation crisis that nevertheless affected the airplane division. The Model T automobile was fast becoming outmoded, and Henry Ford would not admit it. For years the Model T had been the world's most popular automobile, but in the mid-1920s General Motors and Chrysler began to make inroads. It was time of great prosperity, and the car-buying public wanted something more than a good, reliable, inexpensive automobile—they wanted their cars to be sporty and fashionable as well. Model Ts were black and functional. The cars of Ford's competitors were manufactured in all colors, with all types of accessories, and with more rakish lines than the square, box-shaped Model T. When Ford sales dealers tried to persuade Henry Ford to join the trend, he made the famous statement, "A customer can have any color as long as it is black." Sales of Model Ts dropped alarmingly. To try to stem the tide, Ford decided to reduce prices on the various Model T models. Roadsters, for example, were lowered in price to as little as $260. But still sales did not pick up. Frantically imploring Ford to take more decisive action, dealers finally got the auto magnate to bring out the Model T in different colors. But sales continued to sag. Ford dealers complained more loudly. They wanted a new car.

Henry Ford finally gave in. On his sixty-fourth birthday, at his usual press interview, for the first time in years he said nothing about aviation. Now the push was on a new Ford car. "I have the biggest job of my life ahead of me," Ford admitted. "I have to come up with a new model automobile."

For the next four months Ford shut himself up in his laboratory and worked on the design for the new Ford car. Aviation and all his other activities received almost none of his attention. By mid-1927, Ford was ready. When the fifteen-millionth Model T rolled off the production line, manufacturing was halted, and the factory changed over to produce the new Model A car. The first new Ford automobile to appear in more than a decade, it officially appeared in 1928 and proved an outstanding success. With a standard gearshift, low-slung chassis, and new engine capable of driving it 60 to 70 mph, the Model A was eagerly sought by car buyers. "The Tin Lizzie is succeeded by the Elizabeth," newspapers commented, and for weeks Ford showrooms were mobbed.

The success of the Model A encouraged Ford to return with fresh enthusiasm to a project he had long had in mind for the Ford Tri-Motor. He wanted to introduce a new model.

Chapter Eight
Transcontinental Air Service Begins

GSHIP OF TH~~T.A.T.~~ FLEET OF FORD
MONOPLANES OPERATING WI... PENNSYLVANIA AND SANTA FE RAILROADS IN THE NEW R...
G SUNDAY JULY 7...

ACROSS THE CO...

167

Through the years various modifications had been made to improve the 4-AT's performance and accommodations. As the Wright Aeronautical Corporation had improved its Whirlwind engines, so had the engines on the 4-ATs been upgraded. The first Wright Whirlwind engine, the J-4, delivered 200 hp, giving the trimotor a total power of 600 hp. The next Wright engine was a J-5, rated at 220 hp, and after that there was a J-6 engine and a J 6-9 engine, each of which delivered 300 hp. Trimotors using these engines had a total power of 900 hp as compared to the 600 hp of the first 4-ATs.

With these improved engines, 4-ATs were able to fly at faster speeds and carry bigger payloads, but the Fords realized that, for their new model Ford Tri-Motor, they should switch to the new Pratt & Whitney Wasp engines that Boeing was using. By doing so, they would have a speedier and even more powerful Ford Tri-Motor. It was Lindbergh who first had suggested they look into the new Pratt & Whitney engines. The day he visited Ford Airport he told the Fords, "These Wasp engines can develop as much as 420 to 450 hp. If you use them, you can have a total power of more than 1,300 hp."

The Fords took Lindbergh's advice and designed their new Ford Tri-Motor around the Pratt & Whitney engine. The result was the 5-AT, bigger and faster than the 4-AT and able to carry a heavier payload.

The maximum speed and the cruising speed of the 4-AT were 130 mph and 107 mph respectively. The same speeds for the 5-AT were 135 mph and 112 mph. The 4-AT weighed 6,500 pounds and could carry a payload of

Preceding Page: Transcontinental passenger air service starts up on July 7, 1929, at the opening ceremonies in New York's Pennsylvania Station. Christening the T.A.T. Ford Tri-Motor flagship is famed aviatrix Amelia Earhart. Looking on (left) is Grover Whalen, New York's Police Commissioner and official greeter. TRANS WORLD AIRLINES

1,725 pounds. The 5-AT weighed a thousand pounds more (7,500 pounds), and could carry almost double the payload—3,743 pounds.

The 5-AT was larger, too. With a fuselage length of 77 feet 10 inches, it was more than 3 1/2 feet longer than the 4-AT. As a result, it had a longer passenger cabin (18 feet 9 inches in length as compared to 16 feet 3 inches) and could carry fifteen passengers instead of the eleven carried by the 4-AT.

The price of the 5-AT was also higher, chiefly because the Pratt & Whitney engines were more expensive. A customer had to pay $65,000 for a 5-AT as compared to $50,000 for a 4-AT. Since many airlines could not afford the more costly 5-AT trimotor, Ford continued to manufacture the less expensive 4-AT. But one of the Airplane Manufacturing Division's two production lines were turned over to 5-AT production.

Pratt & Whitney itself purchased the first 5-AT made, and used it as a company plane. Other purchasers in the first six months of production included Stout Air Services, Robertson Aircraft, Cia Mexicana de Avacion, Pan American Airways, and Maddux Air Lines (it bought five).

The largest single purchaser of 5-ATs in these early months was an important new airline which started up using Ford Tri-Motors exclusively. This was Transcontinental Air Transport, a new company promoted into existence by the indefatigable C. M. Keys, the Wall Street financier who earlier had organized National Air Transport. Keys had always thought of NAT as a New York-Chicago freight service, despite the fact that its airmail contracts first set it on a Chicago-Dallas route. With Transcontinental Air Transport, Keys had an entirely different concept. This company he saw as a passenger airline, and one that would fly coast to coast. At the time, passenger flying at night had not yet been attempted on any large scale, but Keys had an answer for this. TAT's coast-to-coast passengers would travel by air during the daytime, and by Pullman at night.

Keys was able to interest the Pennsylvania Railroad into becoming the chief backer of the plan; supporting Transcontinental Air Transport to a smaller extent was the Santa Fe Railroad. To serve as the new company's technical adviser, Keys turned—as Juan Trippe of Pan American Airways had done—to Lindbergh. With the inducement of a healthy block of stock,

Lindbergh agreed to take the job of getting the operation underway. He not only plotted the route, but also inspected the airports and weather stations along the way and set up standards for equipment and personnel. Lindbergh became so closely associated with TAT that the airline was known for years as the "Lindbergh Line." Lindbergh was paid handsomely for use of his name, and for years TAT/TWA used the "Lindbergh Line" as its motto, painting it on some of its first DC-2's.

While Lindbergh was getting ready to launch the coast-to-coast service, TAT ordered its fleet of Ford Tri-Motors: the airline purchased ten 5-ATs for a total price of $650,000. It was the largest single order ever placed for commercial air-planes up to that time.

On the heels of this sales coup came more news that reaffirmed the position of Ford Motor Company as the nation's most important aviation leader. It had been exactly three years since the establishment of Ford Air Transport Service, and the annual release of the company's airline's operational statistics revealed an unmatched record of safety and service. In an era when airplane crashes were still frequent, the Ford airline had never had a fatal accident. In a time when airlines were plagued with incomplete flights and disrupted schedules, Ford Air Transport Service had completed 91.1 percent of its flights. In three years of operations, it had flown 1,060,170 miles and carried 6,444,073 pounds of freight in addition to 32,031 pounds of mail. Commented *The New York Times* in an editorial: "This is evidence that the airplane has proved its value and reliability as a freight carrier."

In the summer of 1928, Henry Ford for the first time consolidated some of his aviation operations. On July 19, 1928, without ever giving a public explanation, he discontinued the Ford Air Transport Service's Cleveland-Detroit run, and on the same day he announced that Ford was giving up its airmail contracts. "Other airlines have now developed to the point they can take over the field in which Ford planes have pioneered," Ford explained in a letter to the postmaster of Detroit.

Ford at this time also halted his project to build a "Flivver of the Air." There were two reasons for this decision: one, the realization that such a tiny plane was impractical; the other, the tragic death of Harry Brooks, killed while trying for a record in one of the midget planes. In February, 1928, Brooks had gone to Melbourne, Florida, to test a new version of the plane with a souped-up engine. He started out flying over the ocean in a series of spectacular low dives. But, circling low with the intention of coming in and landing, he hit a high swell with one wing, and the plane fell into the sea. Brooks's body was never recovered.

When interviewed afterward, Ford was obviously very upset. As the son of one of his few personal friends, Brooks was almost like a son himself. "He was a brave and brilliant man," Ford said, adding that his faith was unshaken in the small flivver plane. "We still intend to mass-produce it at a price everyone can afford." But the tragedy left a mark on Ford. Except for one more brief flirtation with the flivver plane a few years later, the company stopped experimenting in this field.

One aviation activity in which Ford's interest did not slacken was the National Air Tour, which in 1928 was bigger than ever. Scheduled to begin June 30 and to last for one month, the tour was to cover 6,300 miles and fifteen states. The planes would start in Dearborn as usual, and then fly as far south as Texas, then west to California, then up to Seattle, where they would turn eastward and head back to Detroit.

Twenty-four planes were to fly in the 1928 National Air Tour. The Ford entry was a 5-AT piloted by Frank M. Hawks. The heaviest of all the planes that would be touring, this Ford Tri-Motor had a total weight, including passengers, freight, and the plane itself, of more than five tons. Other participants in the tour included Eddie Stinson, the previous year's winner, who would again be flying a Stinson Detroiter, Louis Meister in a Buhl Air Sedan, William Brock in a Bellanca, and John P. Wood in a Wright-powered Waco biplane. For the first time there would also be a woman pilot. She was Miss Phoebe Fairgrave Omile, flying a monocoupe.

On opening day, despite a heavy rain, fifty thousand spectators were on hand at Ford Airport to see the tour get underway. The Ford Band played, and, as in previous years, army pilots from nearby Selfridge Field flew in to launch the event with aerial stunts. At 1:08 P.M. Edsel dropped the starter's flag, and the first plane—Miss Omile's—started down the field, splashing through the puddles until it reached the end of the concrete runway. Then, picking up speed, it climbed into the air, circled the field, and headed south toward the first stop, Indianapolis.

The other planes followed, and for the next twenty-eight days they hedgehopped across the western United States. The weather was variable and often rough, but there were no accidents en route. The winner was John P. Wood in his Waco biplane. Frank Hawks and the 5-AT Ford Tri-Motor placed second. Miss Omile, who had the goal of "not winning, but just finishing," did exactly that. Although she had several minor mishaps en route, she did make it back to Dearborn where she straggled in—in last place.

1929 was the biggest year yet for the manufacture and sale of Ford Tri-Motors. During the year, eighty-nine planes were delivered to customers, an increase in sales of more than thirty planes over the previous year. Of the Ford Tri-Motors sold in 1929, about two thirds were the new 5-AT models, the rest were 4-ATs.

As Ford had predicted, production rates had been increased. Now it was possible for the Airplane Manufacturing Division to manufacture as many as three planes a week. To accomplish this, Ford had expanded the factory greatly. Its original number of two hundred employees had been increased to twelve hundred. The factory's floor space was 155 percent greater than it had been a year before.

Such increased production capability enabled Ford to bring down the selling price of the trimotors. He had long promised to reduce the price once he got mass-production underway. Now he reduced 4-ATs from $50,000 to $42,000 and 5-ATs from $65,000 to $55,000.

According to his competition, Ford's reason for the price reduction was not completely altruistic. Fokker, it seems, had been improving his wood and veneer trimotors, and he also was selling planes in larger quantities than before. To meet his orders, Fokker had built a second American factory at Wheeling, West Virginia, and he was looking forward to even more sales, since the prices of his trimotors were lower than those of the Ford Tri-Motors.

With the Ford planes reduced in price, however, some of Fokker's customers switched. A southern airline owner named Erie P. Haliburton ordered five Ford Tri-Motors to be delivered in 1930. He said that up to the time of the price reduction he had been considering Fokkers.

Once the 5-ATs were in full production, the orders for 4-ATs fell off. Those most interested now in 4-ATs were smaller airlines, or corporations desiring company planes. Among the former group were Mamer Flying Service, Spokane; Eastern Air Transport, Brooklyn; and Concesionaires de Line as Aereas Subvencionadas S.A., Madrid (Spain's government-subsidized airline). Among the latter group purchasing 4-ATs were Firestone Tire & Rubber Company and Curtis Publishing Company.

In addition, the armed services kept buying 4-ATs for use as transports. These planes destined for the military were specially fitted, and in the service they were given different designations. The army called them C-3s, the navy and Marine Corps, JR-2s.

But in June, 1929, Ford called a halt to 4-AT production. With the exception of one other 4-AT which was delivered to Concesionaires de Lineas Aereas Subvencionadas S.A. in 1931, no more 4-ATs were ever built. In all, starting from the first 4-AT built in June, 1926, seventy-eight had been manufactured.

The reason for the discontinuance of the 4-AT was that just about everyone wanting a Ford Tri-Motor now preferred the larger and more powerful 5-AT. The U. S. Army which had purchased eight 4-ATs now switched to 5-ATs, purchasing five of these, which they put into the Army Air Transport Service. Designated as C-4 cargo planes, most were used as commerce transports for VIPs such as the Secretary of War.

The 5-ATs sold to the navy and Marine Corps were designated JR-4s, and were modified for use as flying ambulances and paratroop carriers as well as cargo planes.

One big customer for 5-AT Ford Tri-Motors was Pan American Airways, which in 1929 purchased six of the planes. So fast was this airline expanding that hardly a week passed without Pan American's adding several hundred miles of air routes. Pan American was not only pushing down the eastern coast of South America, but via a new airline—Pan American-Grace, which it formed jointly with Grace Steamship Company—it was also organizing an air service down South America's western coast. By the end of the year, Pan American would be operating in twenty-eight countries.

By mid-1929 Transcontinental Air Transport had its fleet of ten Ford Tri-Motors delivered, and it was ready to launch the nation's first transcontinental air passenger service. Something of a service had existed previously, when two other airlines flew passengers—two at a time—over a circuitous route from New York to San Francisco that involved fifteen stops and a time-period almost as long as the railroad's one hundred hours. But Transcontinental Air Transport's service was to be quite different. Nothing as deluxe had ever been seen in American aviation before, not even in Stout Air Services. From the time a passenger started off from either coast, he would be going in style. He would be met at airports by teardrop-shaped "Aero" automobiles designed especially for the airline. He would receive gold pens as flight souvenirs. Aboard the plane he would dine using gold-plated flat wear.

Contributing heavily to the sense of luxury were the Ford Tri-Motors themselves, the only planes flying the TAT route. An extensive advertising campaign calling attention to the inauguration of the service emphasized the performance of the Ford Tri-Motors. "More stable than a yacht," the ads said. "Swifter than the wind . . . a new exaltation . . . the thrill of an indescribable experience . . . complete, luxurious relaxation."

Despite the fact that the fare ($351.94 one way) was high, more than a thousand applications were received for the inaugural flight. Only ten passengers were accepted from each coast. The 5-ATs could fly at least sixteen passengers each, but, to afford greater leg room and comfort on the first flight, it was decided to limit the number of passengers.

Although everyone somehow thought of the trip as being all-air, it was actually as C. M. Keys had planned it: part air and part rail. During the day, the passengers would fly through the air in Ford Tri-Motors. At night they would roll over rails in Pullmans. In the late afternoon of July 7, 1929, the passengers crossing the nation from east to west assembled in Pennsylvania Station, New York City. Here they would board a special train that would take them to Columbus, Ohio, where they would pick up the Ford Tri-Motor that would fly them to Waynoka, Oklahoma. From here they would board a Pullman again, going by the Santa Fe Railroad to Clovis, New Mexico, where again they would transfer to a Ford Tri-Motor for the final hop to Los Angeles, where they would land at Glendale, outside the city.

Pennsylvania Station was thronged with people that afternoon. In one corner of the station a brand-new Ford Tri-Motor stood on display. This plane, the *City of New York,* was, as a sign by it indicated, the "Flagship of the T.A.T. Fleet of Ford Ail-Metal Planes."

Besides the big corrugated-metal plane a wooden speaker's platform had been erected. Here notables had gathered in honor of the first flight. Among them were Grover Whalen, who had left Wanamaker's to become New York City's police commissioner; W. W. Atterbury, president of the Pennsylvania Railroad; Betty Brainerd, a well-known newspaper woman; and noted aviatrix Amelia Earhart, who in a few months would become a TAT executive in charge of a program to make air travel more attractive to women.

Radio station WOR covered the event with a microphone hookup, and the ceremonies were brief. W. W. Atterbury said a few words, comparing the event to the 1889 Golden Spike ceremony marking the linking of the first transcontinental rail system. Amelia Earhart spoke next. She said she was flying on the inaugural flight and carrying letters from James J. Walker, mayor of New York, to the mayors of Los Angeles and San Francisco. The passengers for the inaugural flight had collected in front of the platform. After the speeches were over, they marched in a group to the waiting *Broadway Limited,* which had been renamed the *Airway Limited.*

Meanwhile, on the West Coast, Lindbergh sat in the office of California's governor, C. C. Young. A telegraph key on the governor's desk was connected by Western Union to Pennsylvania Station in New York. Lindbergh looked at his wrist watch, and at 6:05 P.M. (EST) pressed the key. In Pennsylvania Station, the signal was heard, and the *Airway Limited*'s engineer gave three toots on the train's steam whistle. The train pulled slowly out of the station as the Sunrise Trail Band of the Long Island Railroad struck up the tune, "California, Here I Come."

The City of Columbus, *one of the new 5-AT Ford Tri-Motors. Flagship of Transcontinental Air Transport, Inc. (now TWA), it inaugurated transcontinental passenger air service in the United States in 1929.* TRANS WORLD AIRLINES

The next morning the *Airways Limited* pulled into a special depot built at Port Columbus, site of Columbus's new municipal airport. Despite a drizzling rain, five thousand people were there, including Henry and Edsel Ford. A shiny new Ford Tri-Motor, the *City of Columbus,* was waiting. Disembarking from their Pullman, the passengers boarded the plane. For a few minutes, nothing was heard but "ohs" and "ahs." No one had ever seen such luxurious accommodations in an airplane. The reclining seats, the windows that could be opened and shut—this was even better than the Pullman.

Once the passengers were aboard, the pilots started up the Pratt & Whitney engines. The noise was overpowering and the vibration, even worse. A "courier" walked down the aisle handing out little pieces of cotton, which the passengers were told to put into their ears.

Now another telegraph key was pressed, in Washington, D.C., by Secretary of Commerce Robert P. Lamont, A gong sounded at the Port Columbus airport, and the pilots eased off the *City of Columbus's* brakes, sending the plane taxiing down the runway.

Accommodations aboard the T.A.T. Ford Tri-Motors were luxurious for the day. Each passenger had his own window through which to enjoy the View. FORD MOTOR COMPANY

In seconds, the *City of Columbus* was airborne. Climbing about 4,000 feet, the big plane headed southwest, flying at about 100 mph. Since no heaters had yet been installed in 5-ATs, the plane's cabin soon got chilly. But passengers got somewhat warmed up when the courier came down the aisle again and served hot tea on lavender place mats. It was a long way to Waynoka, Oklahoma, but intermediate stops were made at Indianapolis, Saint Louis, Kansas City, and Wichita. After the *City of Columbus* left Saint Louis, lunch was served, the courier bringing out the gold-plated flatware and serving hot consomme, cold cuts, and strawberry shortcake.

The shortcake had been donated by the mayor of Saint Louis, and a little card handed to the passengers informed them of this gift.

Lindbergh was now at the Grand Central Air Terminal in Glendale, California, making ready to fly out the West Coast passengers in the *City of Los Angeles*. According to one account, Lindbergh to practice up before the actual flight had gotten to the airfield at 7 A.M. and had taken the *City of Los Angeles* up and landed it ten times, all in the course of two hours.

By the time of the actual flight later that morning, thirty thousand people had gathered at the Grand Central Air Terminal. Screen star Mary Pickford christened the *City of Los Angeles* with a bottle of grape juice (Prohibition forbade any alcoholic beverages). Then the passengers climbed aboard. One was Anne Morrow Lindbergh, a bride of only two months.

Lindbergh himself piloted the *City of Los Angeles* on the takeoff, and after they had flown eastward for about an hour, he walked back into the passenger section. Explaining some of the plane's technical features to the passengers, he mentioned that the *City of Los Angeles* was equipped with a plane-to-ground radio, and he offered to transmit any commercial messages. A Los Angeles newspaperman named Gene Coughlin said he wanted to radio his city editor for fifty dollars.

"That's not a commercial message," Lindbergh said.

"Colonel, that's the most commercial message you'll ever see," Coughlin replied.

Lindberg sent the message.

When the East Coast passengers in the *City of Columbus* landed in Waynoka, Oklahoma, they were whisked by the special "Aero" automobiles to the Santa Fe station where a Pullman car waited to take them on an overnight trip to Clovis, New Mexico. Here they boarded another Ford Tri-Motor that would fly them to Grand Central Air Terminal in Glendale.

Their pilot was Lindbergh, who had arrived earlier with the West Coast passengers. Another huge crowd was waiting at Glendale, and this time screen star Gloria Swanson was on hand with the bottle of grape juice, ready to christen the Ford Tri-Motor arriving from the east. Also on hand were four Fox Movietone newsreel cameramen. They would record the event for one of the first "talkie" newsreels. A big cheer resounded when Lindbergh brought the plane in and touched its wheels on the airfield's runways. The passengers climbed out, praising the trip and the service they had received. "I have never felt more comfortable and less wearied," one said.

Newspapers heralded the event. The coast-to-coast passengers had flown 2,343 miles by air, and 970 miles by rail. In forty-eight hours they had crossed the continent, slightly less than half the time it took by rail alone.

Other airlines were also making news. Stout Air Services, the oldest and best-known passenger airline in the United States, had continued to expand its services. By now it was flying scheduled flights not only to Cleveland and Chicago, but also to Battle Creek, Kalamazoo, and South Bend. As Ford Air Transport Service had established records in shipping freight by air, so had Stout Air Services in the area of passenger transportation. In three years of operation, Stout Air Services had carried almost two hundred thousand passengers, and had never had a fatality.

The success of Stout Air Services was not unknown to the organizers of a new giant aviation company. For once the guiding spirit was not C. M. Keys, the Wall Street financier. This time it was Frederick Rentschler of Pratt & Whitney and Bill Boeing of Boeing Air Transport. Taking a leaf from Keys's book, they decided to organize an aviation holding company into which they would combine many other companies—those that manufactured airplanes and parts, and those that operated airlines.

The name of the new company was United Aircraft & Transport Corporation, and as a starter it contained Boeing Airplane Company, Boeing Air Transport, Pacific Air Transport, and Pratt & Whitney. Among other companies quickly added were Stearman Aircraft, which made small planes; Northrop Aircraft, manufacturers of military trainers; Sikorsky Aircraft, which specialized in seaplanes and amphibians; and Chance Vought, a company that produced fast navy planes.

As part of its expansion program, United Aircraft & Transport Corporation began thinking about establishing its own transcontinental service to compete with TAT's. One of the United companies—Boeing—was already flying passengers and freight between San Francisco and Chicago. To fill in the remaining link between Chicago and New York, United had its eye on National Air Transport, the company that Keys had founded. But NAT was primarily a freight line. It was still using planes with open cockpits, and could carry no more than one or two passengers on most flights.

United realized that if it wanted to add a passenger service to connect with Boeing in Chicago, it had two choices: 1) to organize a new passenger airline, or 2) to buy up already existing passenger airlines along the New York-Chicago route. The latter course seemed far more preferable, and this is when United began to look more and more in the direction of Stout Air Services. Because Stout's airline flew from Chicago to Cleveland, the acquisition of this airline would give United more than half the linkage it desired between Chicago and New York. Besides that, Stout Air Services had other attractions. It owned an entire fleet of Ford Tri-Motors. Also, its personnel were experienced in handling passengers.

Rentschler, the president of United, went to see Stout. The eccentric inventor realized that he had taken his airline as far as he could. Although Stout himself had had the dream of a big national airline, Stout Air Services had neither the capital nor the big-business backing to extend its services further.

United on the other hand was big enough to make a national airline a reality.

After thinking it over, Stout decided it was best to sell out. His stockholders, many of whom had been with him since the days of the Stout Metal Airplane Company, backed his decision. Although they had been promised they would never see again their original investment of $1,000, they saw it doubled once more with the United offer. In all, for every $1,000 invested, $4,000 had been received as a result of the first Ford take-over, and now United's.

In August, 1929, the stock of Stout Air Services was officially transferred to United, and Stout made his exit from commercial aviation. He still was a Ford vice-president and a director of the Airplane Manufacturing Division.

He still too would continue his aeronautical experiments. (Indeed, he continued these until his death in the late 1950s.) But never again would he be in a position to make a major contribution to commercial aviation.

Henry Ford however was still very much involved in aviation, and in mid-1929 he issued, as he did every year, his annual statistics on the operations of Ford Air Transport Service. This particular year his announcement stressed the increased reliability of trimotor planes. The Ford private airline was still operating a few single-engined 2-ATs, and as these were replaced, one at a time, by Ford Tri-Motor 4-ATs and 5-ATs, Ford noted that the airline's record improved. "Single-engined planes were involved in most of the delays and 90 percent of the forced landings," he said.

As usual, the Ford Air Transport Service's report produced a favorable response from the press. Commented a *New York Times* editorial: "These statistics give us the best picture available in this country of the commercial use of the airplane over a long period."

Just before the historic first flight over the South Pole, Byrd weights an American flag with a stone from Floyd Bennett's grave. Flags were also dropped over the Pole for Amundsen and Scott.
CULVER PICTURES

Another aviation activity involving Ford was also making headlines. Byrd was about to attempt his flight over the South Pole in a Ford Tri-Motor. In the spring of 1928 Byrd was completing plans for the expedition. He expected to spend two years in Antarctica. Once he reached the south polar region he would set up a base, and from here he would conduct his explorations and attempt his South Pole flight.

Besides the Ford Tri-Motor, which the Fords had donated, Byrd would take three other airplanes on the expedition: a Fokker Universal, a General Aircraft plane, and a Fairchild cabin plane to be used for photographic reconnaissance. The Ford Tri-Motor was a brand-new 4-AT. Built in March, 1928, it had been refitted in Dearborn for the expedition. A heavy-duty engine had been installed on the nose, and the plane was also adapted so skis could be attached for taking off and landing on ice. Byrd named the Ford Tri-Motor the *Floyd Bennett* in memory of the pilot who had flown with him over the North Pole.

Bad luck had dogged Bennett ever since their memorable flight in the *Josephine Ford.* He was to be the pilot on Byrd's transatlantic flight in the *America,* but was seriously injured testing the plane. By the time the Antarctic expedition was almost ready to sail, he had recovered. While waiting for preparations to be completed, he had volunteered to fly to the rescue of two German aviators forced down in the Guff of St. Lawrence. The rescue flight was successful, but as a result of it, Bennett contracted pneumonia. He died in Quebec in April, 1928.

Five months later, without Bennett, the big Ford Tri-Motor was tested at Mitchel Field in Long Island. The plane's pilot now was the veteran Bernt Balchen. With two mechanics aboard as well as Harold I. June, who would be the radio operator on the South Pole flight, Balchen took the big plane up. Its total weight was 13,000 pounds, but the *Floyd Bennett* was able to climb to ten thousand feet without difficulty. Byrd; watching from the ground, was pleased.

After the other planes were tested, Byrd had them all flown to Hampton Roads, Virginia, where they were dismantled and stowed aboard the expedition's ships. Once they were loaded, the ships traveled through the Panama Canal to San Pedro, California. Here more men and supplies would be taken aboard, and they would set sail on the long sea voyage to Antarctica.

The Floyd Bennett *being loaded aboard the whaler* Larsen *at Hampton Roads, Virginia, in August, 1928.* PICTURE COLLECTION, NEW YORK PUBLIC LIBRARY

Byrd established his base, Little America, on the Ross Shelf Ice, south of New Zealand, near 165° W. From here many successful short flights were made in the *Floyd Bennett,* all leading up to the big polar flight. As the weeks passed, Byrd laid his plans carefully. To minimize the risk, parties had been sent by land as far as possible over the treacherous ice and snow to establish a line of supply depots on the direct route to the South Pole. In all, eight depots had been set up. If the big Ford Tri-Motor came down unexpectedly, the crew would at least have a chance to locate some food.

When all was ready, everyone waited anxiously for the weather to clear. The *Floyd Bennett,* now equipped with skis, was stowed in an ice hangar. Finally, on Thanksgiving Day, November 25, 1929, an advance meteorological unit radioed back the message: VISIBILITY CLEAR. A dogsled team of sixty Alaskan huskies pulled the *Floyd Bennett* out of the hangar onto runways cut into the ice. The entire expedition was assembled outside, all dressed in heavy Eskimo parkas. Once the big plane was eased into position, the four men scheduled to make the flight prepared to go aboard. These were Byrd, who would serve as navigator; Balchen, the pilot; June, the radio operator; and Captain A. C. McKinley, the map photographer.

The radio gear and other equipment had already been stowed inside the *Floyd Bennett.* Now large sacks of food and extra cans of gasoline were loaded aboard. The four men climbed in, Balchen and Byrd slipping into the pilots' bucket seats. Checking the instruments and gauges on the dashboard, Balchen ordered the plane's engines started. Sputtering at first, the engines soon roared lustily. Byrd then gave the signal, and as the expedition members waved goodbye, the plane started down the runway. With a gross weight of more than six tons, the plane had trouble getting airborne, but, its engines straining, it slowly lifted itself into the air and at 3:15 P.M. was heading southward toward the South Pole, more than five hundred dangerous miles away.

The assignment Byrd and his men had chosen for themselves was staggering. If everything went all right, there would be no problems. But they were flying over the world's worst terrain, and they were the first ever to fly over it. Any number of unforeseen perils lay ahead. They had taken many precautions. They had placed supply depots of food along part of their route. They also carried a month's supply of food aboard the *Floyd Bennett.* But all this depended on being able to land the plane safely in case they were forced down. Such a landing might be impossible.

After leaving Little America, the *Floyd Bennett* picked up a tail wind, and for about six hours no problems were encountered. Actually Byrd had expected none along this part of the route. Ahead of them was something different, however. Blocking their path to the South Pole was a high, dangerous mountain range. Through it ran a pass previously traversed only on foot by Amundsen and Scott. Years before, these explorers had tortuously inched their way over the bitter-cold, treacherous terrain, taking weeks to get through. Byrd—if the *Floyd Bennett* could get through at all—would fly through in minutes.

Its three engines throbbing mightily, the *Floyd Bennett* kept flying southward at about 100 mph. Realizing they would soon come to the difficult mountain range, Balchen tried climbing as high as he could. At 9:15 A.M., they spotted in the distance the mountain range barricading their way to the South Pole. Balchen tried to climb. The plane responded sluggishly. Slowly the needle on the altimeter moved. Five thousand feet. Seven thousand feet. They were nearing the mountains now, and Byrd suddenly pointed. There was the pass through the jagged, ice-cragged mountains.

They needed much more altitude. Balchen tried again to climb. The overloaded plane strained, and it did climb slightly—8,200 feet, the altimeter read, 9,400 feet. They had reached the pass now, and Balchen nosed the plane in. Mountains loomed on both sides. The pass was narrower than they had thought. There was no room to turn the plane around. They had to keep going now.

Byrd pointed anxiously. At the end of the pass was a high ridge that had to be cleared.

The Floyd Bennett *is chipped out of the ice upon Byrd's return to Antarctica in 1933. After minor repairs, it soon flew perfectly,* FORD ARCHIVES PHOTO

Balchen tried climbing again, but the *Floyd Bennett* could rise no higher. "We've got to lighten the ship," Balchen shouted.

Byrd hesitated. Should they throw out some of the precious food rations, or should it be the extra gasoline they might need for the return trip to Little America?

Byrd made his decision. "Dump out four days' food supply."

The plane door was quickly opened and the food sacks pushed out. Shuddering, the plane lifted slightly, but down-drafts sweeping through the pass pushed it back again.

"We're stalling!" Balchen shouted. "We've got to lighten the ship more!"

Byrd nodded. "Throw out more food sacks," he called to June and McKinley.

Again the *Floyd Bennett* shuddered, and for a few seconds, nothing happened. Byrd looked with alarm. They were now nearing the ridge at the edge of the pass. Suddenly, just when they reached the top of the ridge, the

plane clawed up an extra three hundred feet—and, with inches to spare, cleared the pass.

Before the *Floyd Bennett* now was a vast glacier plateau. Smooth and unbroken, it stretched straight to the South Pole. The plane, its engines throbbing mightily, swept across the plateau. At 1:15 A.M. Byrd shot a navigational fix and then excitedly tapped Balchen, telling him to circle.

"Ninety, south," he yelled back to the other two men in the cabin.

Everyone pounded everyone else on the shoulder. They had made it! Directly below was the South Pole. By slogging over snow and ice, Amundsen and Scott had taken five months to reach the South Pole. The *Floyd Bennett* flew it in ten hours.

As Balchen kept circling the South Pole, Byrd dropped three American flags. The first, in honor of Floyd Bennett, was weighted with a stone from Bennett's grave. The next two were for Amundsen and Scott. Then, checking his sun compass once again, Byrd directed Balchen to start back. Heading northward, the *Floyd Bennett* cleared the pass, this time without any difficulty. They returned to Little America nineteen hours after they had started out. In all they had flown 1,135 miles.

In honor of the great achievement, Byrd a month later was made a Rear Admiral. A year later, in 1930, the expedition returned to the United States. The *Floyd Bennett* was left at Little America, exposed to the ice and snow. In 1933, Byrd returned at the head of a second Antarctic expedition. They found the *Floyd Bennett* where they had left it, but now it was completely enclosed in ice. Chipping out the big plane, they punched some holes in the plane's fuselage, but otherwise the *Floyd Bennett* was intact. Under Byrd's supervision, its carburetors were cleaned, new batteries installed, and the plane refueled. Then came the big moment when they went to start the engines. With a sputter, the propellers turned and then spun rapidly, the plane's three engines churning the air as they roared.

A few days later Byrd took up the *Floyd Bennett* again, and when he returned to the United States, he brought the historic Ford Tri-Motor back with him. Today it can be seen at the Henry Ford Museum in Greenfield Village, Dearborn, Michigan.

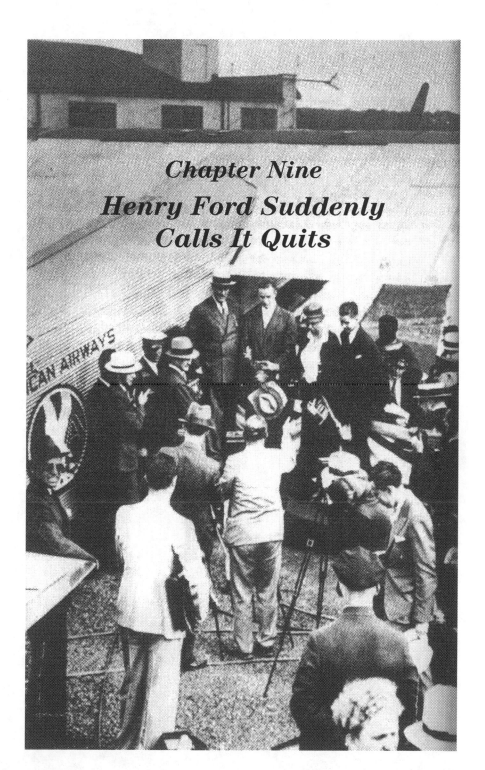

Chapter Nine
Henry Ford Suddenly Calls It Quits

Since World War I the economy of the United States had been increasingly bullish each successive year. Millions of Americans, blue-collar as well as white-collar, were speculating on the stock market, and were more involved with their securities than with their jobs, their businesses, and even their families.

And then the bubble burst.

Throughout the summer of 1929 dangerous signs appeared, but almost everyone, from President Hoover in Washington, D.C., to the House of Morgan in New York City, ignored them. By October 29, 1929, it was too late. This was Black Tuesday, the day the stock market took the disastrous plunge that catapulted the nation and the rest of the world into the Great Depression. The once highly touted aviation stocks collapsed with the rest of the market, and the airlines entered a dark age in which they would flounder for many years.

The fact that the major airlines were able to survive at all can be largely attributed to the Watres Act, which gave the industry its biggest legislative boost since the Kelly Act of five years before. This act was the brainchild of Walter Folger Brown, the postmaster general appointed by President Herbert Hoover in 1929. An attorney from Toledo, Brown had been an assistant to Hoover when Hoover had been secretary of commerce, and he had ended up handling Hoover's political affairs. Everyone thought Brown's appointment a political payoff, and few expected him to become involved in anything controversial or constructive.

But Brown had a grand plan for the nation's airlines, not so much to save them from the effects of the depression, but rather to elevate them to a level that no one had anticipated before. In essence, Brown wanted

Opposite: On July 2, 1932, in Albany, New York, Franklin Delano Roosevelt boarded a Ford Tri-Motor to fly to Chicago to accept the Democratic nomination for President. AMERICAN AIRLINES

to consolidate commercial aviation by making the big airlines bigger and getting rid of the dozens upon dozens of small companies cluttering up the aviation picture. He also wanted to consolidate the hodgepodge of air routes and to end up with only meaningful routes.

After numerous consultations with congressmen and airline executives, Brown drafted legislation that was submitted and passed on April 29, 1930, under the sponsorship of Congressman Laurence H. Watres of Pennsylvania. The key to the Watres Act was that it offered subsidies to those airlines that increased their facilities and services. Now the federal government would pay for far more than just carrying airmail. Besides this basic rate, airlines would also receive subsidies for the number of passenger seats its planes had, for its two-way radios, for flying at night and over mountains, and for many other "variables."

The provisions of the Watres Act were not lost on the airlines. Now they would get paid for services and space whether they flew any airmail or not. An airline with big passenger planes like Ford Tri-Motors would benefit. Those with open-cockpit planes or planes able to carry only one or two passengers were at a disadvantage.

The Watres Act succeeded beyond all measure in accomplishing its goal, and proof of the accomplishment was the emergence soon of what became today's major United States airlines. Seeing the handwriting on the wall, the smaller independent airlines hastened to sell out their companies, their airplanes, and their routes. The larger airlines were correspondingly only too happy to extend their routes (getting more subsidies) by acquiring the smaller companies.

Transcontinental Air Transport was one of the first airlines to expand as a result of the Watres Act. First it merged with Western Air Express. Then it negotiated to take over Maddux Airlines and its fleet of Ford Tri-Motors. Then it added Standard Airlines. A small line that whisked Hollywood movie stars to weekend retreats in Phoenix and Tucson, this company had grown out of a flying school, stunt flying service, and airplane sales office founded in 1926 by two young pilots named Jack Frye and Paul E. Richter. From these four companies—Transcontinental Air Transport, Western Air

Express, Maddux Airlines, and Standard Airlines—a new company was formed. Its name was Transcontinental Western Air Express, or, as it soon came to be called, TWA.

Transcontinental Air Transport remained as a division of TWA, and less than a month after the merger, it inaugurated—with its same fleet of Ford Tri-Motors—the first all-air transcontinental service. Its original part-air, part-railroad service had been a failure. Newspapers had praised to the skies the accomplishment of the original service, and so much publicity had been generated that Western Air Express had established a similar part-air, part-rail, coast-to-coast service in conjunction with the New York Central Railroad. But financially Transcontinental Air Transport's coast-to-coast service had not paid off. With fares based on rates of sixteen cents per mile, only the wealthiest passengers could afford the trip. Then, too, there was the inconvenience of transferring so many times between trains and planes. When passengers finally arrived after traveling across the continent, they had spent a long forty-eight hours getting there. In eighteen months of operations, TAT and the Pennsylvania Railroad had lost about three million dollars.

Now, with the Watres Act, subsidies would help make up the losses. Except for the fact that the same TAT Ford Tri-Motors would be used, everything else about the new coast-to-coast service would be different. It would be all-air. The time would be shorter—thirty-six hours as compared to forty-eight hours. The price would be less, only $203.50 one way. The route would be different, too. Passengers from the East Coast would depart, not from Pennsylvania Station, but from Newark Airport. The stops would be Philadelphia, Harrisburg, Pittsburgh, Columbus, Indianapolis, Saint Louis, Kansas City, Wichita, Albuquerque, Winslow, and the final destination, Los Angeles. Every stop but Kansas City would be for ten minutes. But at Kansas City, because passenger flights were still not flown at night, there would be an all-night layover. Passengers would be driven by limousines into the city to spend the night at the Muehlebach Hotel.

The inaugural flight took place on October 23, 1930. It was a cold day, and the night before, mechanics at Newark Airport had gone out to the hangars periodically to start up the engines of the Ford Tri-Motors so they would not freeze. Three sections were to be run, two of which would carry passengers and fly Ford Tri-Motors, the other to carry freight and mail and fly a Fokker trimotor.

By 8 A.M., all three planes were on the runway. Assembled to see them off were Postmaster General Walter Folger Brown; Amelia Earhart; Harris M. Hanshue, president of the new TWA company; and J. Cheever Cowdin, a TWA director, later to become president of Universal Pictures. The plan had been for Postmaster General Brown to press a button on the landing field that would electrically start the engines of the *City of Columbus,* the first Ford Tri-Motor scheduled to take off. But the elaborate electrical connection failed. Mechanics had to start the engines the old way by spinning the propellers by hand.

At 8:10 A.M., the *City of Columbus,* with seven passengers aboard, took off, followed fifteen minutes later by the second Ford Tri-Motor, and ten minutes after that by the Fokker. Right on schedule, at 8:16 P.M. the following day, the *City of Columbus* landed at Glendale's Grand Central Air Terminal. All-air transcontinental service had arrived.

By this time United Aircraft & Transport Corporation was filling in the remaining links it needed for its transcontinental service. Since acquiring Stout Air Services, United had taken over other airlines and other routes; but only with difficulty were Rentschler and Boeing able to bring National Air Transport into their corporate fold. C. M. Keys, founder of NAT, had strenuously resisted the take-over, and only after a bitter proxy fight and court litigation did Rentschler and Boeing emerge victorious. Once NAT was acquired, the manufacturing companies in United Aircraft & Transport Corporation—companies like Pratt & Whitney and Boeing Airplane Company—were made separate, and a new company was organized to operate the many airlines in the corporation. Headquartered in Chicago, this new company was given the name of United Airlines.

By 1931 United Airlines had bought out Varney Air Lines and was operating routes from Chicago in all directions. It was not only flying from New York to San Francisco by way of Chicago and Salt Lake City; it had also extended its service southward from Chicago, flying NAT's old route to Dallas. By the end of 1931, it had carried 42,928 passengers, three times the number of the previous year. In 1932, its number of passengers doubled, and the following year it carried an incredible 127,000 passengers.

One other transcontinental route was being forged at this time. This evolved out of another new gigantic holding company called simply Aviation Corporation. Organized in 1929, this company within two years assembled under one corporate roof no fewer than eighty-five different firms. Most of these were airlines, including among them Robertson, Colonial, and Embry-Riddle. But some were entirely different, like radio stations, bus lines, and construction companies. In 1930, spurred on by the Watres Act, a decision was made to remove all the airlines from the corporate conglomerate and bind them together into one airline company. The name of this new company was American Airways, Incorporated, and by linking together the routes of its many different airlines, it too was able to launch a transcontinental service.

Meanwhile, other airlines were also taking advantage of the Watres Act and expanding. Northwest Airlines, which originally flew between the Twin Cities and Chicago, was now flying routes all the way to Seattle, on the West Coast. Eastern Air Transport, which originally flew mail from New York to Washington and then extended its route to Richmond, Virginia, was now flying passengers as well as freight and mail.

Eastern's schedule called for daily flights to be flown weekdays, one flight being made daily in each direction. The complete 310-mile trip from New York to Richmond took four hours. From New York, the flights started at North Beach Airport (now LaGuardia), and stops were made at Camden (for Philadelphia), Baltimore, and Washington.

Eastern's inaugural flight took place on August 18, 1930. The airline used 5-AT Ford Tri-Motors exclusively, and the pilot who flew the trimotor that first day was Captain E. H. "Pete" Parker. Handsome and strapping, Parker showed up at North Beach Airport in a very natty outfit consisting of a sport coat, wool sweater, and light-colored knickers. After posing for photographs, he made sure his fifteen passengers were securely aboard, and then he took off. The day was hot and summery, and everyone aboard, including Parker, opened the trimotor's windows. Parker started to climb, but after he got to two thousand feet, he leveled off because a few hundred feet higher, floated dense clumps of fluffy white cumulus clouds. If he ever got into these, he would be unable to see the Pennsylvania Railroad tracks below that he used as his chief navigational aid.

Staying low, Parker followed this "Iron Compass" for almost one hundred miles. He cross-checked his bearings with a New Jersey road map that he had picked up from a filling station near North Beach Airport. As he flew, he kept the open map across his knees. Once, when a wind came up, the map almost went sailing out the Ford Tri-Motor's window.

At one point where U.S. Highway 1 paralleled the railroad tracks, Parker noticed an automobile below racing his plane. He was so low he could recognize that it was a Hup-mobile. Someone in the car waved a handkerchief. Sticking his hand outside the cockpit window, Parker waved back.

Parker made Richmond on schedule, but on subsequent flights he was not always so lucky. When the wind was right, the Ford Tri-Motor made 90-100 mph, but sometimes there was a stiff headwind, and he could go no faster than 55 mph. Usually Parker piloted the plane the entire run. He always had a copilot, but this man's job was to take the tickets, hand cotton out to the passengers, and tend to any other needs they might have. Small vials of ammonia were also distributed to passengers in case they felt faint. On one trip several weeks after the inaugural flight, Parker's copilot rushed into the pilots' compartment. A little old lady had astounded him by putting the ammonia vials in her ears, and the cotton in her mouth. "She's sitting there with the funniest look on her face," the copilot said.

Flying Eastern Air Transport's New York to Richmond route was not always without its difficulties. Because of the publicity received from the inaugural flights, "passengers" without tickets posed a problem. This was best solved by putting the freeloaders off at the end of the runway.

Another problem was the poor condition of many landing fields along the way. Either all-dirt or covered with cinders, they were rough on the

Ford Tri-Motor's wheels. Often a tail wheel broke, and to play it safe, Parker carried along a spare wheel. If the spare had to be installed, Parker called the passengers out and enlisted them in the job of raising the rear of the airplane. When it was high enough, Parker and his copilot put on the spare.

Sometimes when Parker was flying the south-to-north route, he encountered low-hanging clouds over New York that hid the outlines of the Brooklyn Bridge. Not wanting to chance flying over the bridge, he would fly under it. "The passengers never seemed to complain," Parker said. "I guess they thought we always came in that way."

Of all the United States airlines, the biggest was Pan American Airways, which by 1930 was flying forty thousand miles of air routes and was the largest airline in the world. Having now reached Buenos Aires, far down the east coast of South America, Pan American had also extended its service down South America's western coast with Panagra, the company it owned jointly with Grace. The airline was also beginning to fly to other continents. It purchased an airline in Alaska. It acquired a 45 percent interest in the national airline of China. It flew across the Pacific and established bases on coral islands whose names most of the world had never before heard. On its routes over water, Pan American used Sikorsky flying boats, but for those over land it employed as many as twenty-nine Ford Tri-Motors at one time. Indeed, by now, all major United States airlines—TWA, United, American, Northwest, and Eastern—were flying Ford Tri-Motors. These airlines also advertised their use of Ford Tri-Motors. "Just think," a Northwest Airways advertisement confided, "windows that open and close, and complete lavatory facilities."

Ford itself advertised its planes as being used by various airlines. In July, 1931, in the *National Geographic,* a full-page advertisement featured a photograph of five Ford Tri-Motors lined up on an airfield runway. Not often does United's NAT division assemble its fleet of Ford Tri-Motored, all-metal planes, because, like a railroad, the "rolling stock" must be kept moving. Every hour, day and night, a NAT plane is humming through the skies on its scheduled way, carrying cargo of passengers, mail, or express. . . . Five years' experience in transport flying and eleven million miles of successful operation are the foundation of this necessary transportation service. . . . Ford planes are in demand wherever the American public has learned to accept aviation as a commercial factor of importance. Last year alone, Ford planes flew 8,000,000 miles.

It also ran a special advertising campaign of its own—via full-page ads in the National Geographic and the <u>Saturday Evening Post</u> aimed at the public to inform it there should not be any reluctance now to travel by air, that passenger air travel had now arrived and was as safe and reliable form of transportation as the highway and the railroad.

YOUR PLACE...

YOU CAN NOW accept the mechanical marvel of flight as a fact accomplished. The thrill of flight comes no longer from mere amazement at the performance of a flying machine, but rather in discovering the wide sea above.

During the last ten years man has found a new road to freedom, a safe serene way through the deep blue expanse of the sky. He consults a flying schedule as he does a railroad timetable, and takes his flight comfortably, nonchalantly, no longer an adventurer, but a sensible traveler above the roaring cities, removed immeasurably from the noise and tenseness of the busy world. Swifter than any machine on earth, he outpaces the world below him . . . detached, relaxed, superior to all its petty confusion.

The incomparable charm of the skyways seems to be the heritage of youth. But this is true only because of the natural instinct of mankind to resent a change in habits. There are still old men today who will never ride in automobiles, because they cling to the fears of a plodding generation. Many more will never rise above the earth, because they have become habituated to surface transportation.

When you recall that only ten years ago winged relics of the war were still lumbering precariously overhead, and any man who flew was deemed a hero, doesn't it seem incredible that women are being taught in groups how to fly . . . and that 12,000 paying passengers flew from one airport near New York between dawn and dusk of a single day?

But the greatest progress of all has been made in the improvement of transport planes, commercial carriers.

These planes, carrying from 7 to 14 passengers, are used customarily on the great airlines that streak the skies. They are employed also as private yachts by wealthy men who fly on the wings of the times.

The de luxe Ford plane, all-metal and tri-motored, has made the skyways an acceptable avenue of safe transportation. It has reduced hazards to a reasonable risk, and practically eliminated hardship and discomfort from swift flight. It provides you with a degree of luxury that is comparable to a yacht, and a command over time that is of greatest value to those whose time is limited.

· FORD MOTOR COMPANY ·

One of a series of ten advertisements Ford Motor Co. ran in the Saturday Evening Post and <u>National Geographic</u> to inform the public that passenger air travel in airplanes was now a reality"

Ford Tri-Motors continued to make news in other ways. On September 29, 1930, an experimental 5-AT with special streamlining and controls set a world's record for a heavy transport. The plane carried 4,409 pounds of dead weight over a closed course of 62.3 miles at a record speed of 164.4 mph.

The same year the United States Army started a well-publicized run from Chicago to Seattle. Participating was a squadron of twenty planes, one of which was a Ford Tri-Motor. By this time army planes were flying at speeds far faster than the lumbering trimotors. Taking this into account, the army officer in charge assigned to the Ford Tri-Motor the Cinderella role. Figuring it would lag behind the other planes, he gave it the job of stopping at all the airports en route and picking up the repair work of other planes. But at the end of the third day, when the army planes were scheduled to arrive in Seattle, only the Ford Tri-Motor showed up. The faster army planes had been grounded over Montana by a snowstorm, but the Ford pilot had somehow been able to fly right through the storm.

Even though all-metal planes were now accepted throughout the world as the most safe and most reliable for passenger service, some airlines continued to fly wood and veneer planes like the Fokker trimotor which, its construction notwithstanding, was well designed and performed well. But an end to the use of Fokkers came suddenly after March 31, 1931, when a Fokker trimotor crashed during a thunderstorm near Cotton-wood Falls, Kansas, killing all its passengers, including Knute Rockne, Notre Dame's famous football coach.

An investigation showed that the plane's wing had broken off in the storm, and an examination of the wreckage revealed a structural failure: the Fokker's spars and ribs had begun to rot. Stout's warning to Fokker about wood-constructed airplanes getting "veneer-eal" disease was tragically proven.

Almost as if by magic, the Fokkers disappeared from the major air routes, and the Ford Tri-Motors became an even greater symbol of reliability and safety. Incident after incident was reported in the press, reaffirming the plane's reputation. The record of the Ford Air Transport Service had also contributed greatly to the aura of safety and reliability surrounding

Ford Tri-Motors. By 1931, this private airline of the Ford Motor Company had carried more freight than any other airline in the world, and its value was repeatedly demonstrated by incidents that the Fords never failed to publicize. One day a desperate phone call was received in Dearborn from Ford's Chicago plant. An unforeseen shortage of a particular Model A part had developed. Until the plant received a new shipment of the parts, it would have to temporarily shut down an assembly line. Dearborn saw to it that the parts were quickly loaded aboard the next Ford Air Transport Service plane scheduled to fly to Chicago. Within three hours they were on the floor of the plant. "The expenses that flight avoided paid for our Ford Air Transport Service for many weeks," a Ford executive said.

The Josephine Ford, *the Fokker trimotor donated by Henry and Edsel Ford, in which Richard E. Byrd became the first to fly over the North Pole.* FORD MOTOR COMPANY

Ford Tri-Motors also continued to demonstrate their reliability in the National Air Tours. In the 1930 tour, thirty-five planes participated, and the tour covered more than four thousand miles with stops at twenty-nine cities from Enid, Oklahoma, to Edmonton, Canada. And this year, at long last, a Ford Tri-Motor won. The pilot was Harry Russell, and he won the tour again in 1931 when the entrants made an even longer trip—five thousand miles—from Texas to Ontario. This was the last National Air

Tour. In the seven years it had been run since its earliest days as the Ford Reliability Tour, this promotional tour had accomplished its objective: it had succeeded in getting across to the American public the fact that commercial aviation was reliable and safe. Like the Glidden Tours, which had popularized automobile travel a quarter century before, there was no need to continue them.

Almost every week saw either Henry Ford or a Ford Tri-Motor in the news. In April, 1929, a report was confirmed by the government of the Union of South Africa that it had been negotiating with the Ford Motor Company to establish a commercial passenger and freight airline between the Rand and Capetown. A few months later, the Danish government announced that Ford Motor Company was considering erecting a factory in Denmark to build Ford Tri-Motors, and that some of the planes manufactured there would be used to launch an airline service between Copenhagen and several towns in Jutland. As part of this service, mail and newspapers would be dropped by parachute to various cities along the route.

Nothing developed from these contemplated projects, but another Ford Tri-Motor incident in the United States had great historical importance. On July 1, 1932, Franklin Delano Roosevelt was nominated by the Democratic Convention in Chicago to be its presidential candidate. The then governor of New York searched for a means of opening his campaign in the most colorful, dramatic way possible. Never before had a presidential candidate flown to a convention to accept his nomination. This, Roosevelt decided to do.

The plane chartered for the historic flight was a Ford Tri-Motor. Early that morning it had been flown to the Albany airport. That night, as soon as word was received of the nomination, the plane's pilots were told to warm up the engines. Arriving by motorcade from the Governor's Mansion, FDR boarded the Ford Tri-Motor with vigor and confidence. Accompanying him were James A. Farley, his campaign manager, and four of his children, his daughter, Anna Roosevelt Dall, his older sons James and Franklin, Jr., and his youngest son, John. Taking off immediately, the Ford Tri-Motor flew westward, making only three refueling stops. On the way FDR polished the acceptance speech he would read to the convention delegates.

Notified of his pending arrival by plane, the convention did not adjourn. At 1 A.M. the Ford Tri-Motor landed in Chicago. Welcoming FDR was Chicago's mayor, Anton Cermak, who a few months later would be killed in Miami by an assassin's bullet meant for FDR.

A motorcycle escort whisked Roosevelt to the convention hall, where he was greeted by tumultuous cheering. When the applause died down, FDR began his speech. One of the phrases that he added aboard the Ford Tri-Motor became a keynote of United States policy for the next sixteen years. "I pledge, I pledge myself, to a new deal for the American people."

With Fokkers gone from the major air routes, Ford Tri-Motors were flying most of the nation's air passengers. By and large these Ford Tri-Motors were 5-ATs, the 4-ATs having been shunted to the smaller airlines and the less important routes. Many different varieties of 5-ATs were being flown. From the first 5-ATs manufactured in the summer of 1928, a considerable number of improvements had been made.

Many of these improvements had been added in an effort to increase the speed of the 5-AT. For example, the planes had been further streamlined. On some, pants (enclosures over the tops of the wheels) were installed. On others, engines were mounted inside nacelles (metal skins forming podlike shells around the engines). On still others, landing gears had been "faired" (streamlined). With all these improvements, as much as 12 mph was added to the speed of some 5-ATs.

In other efforts to increase speed, modifications were made in power plants and propellers. Some planes were built with two-bladed propellers, others with three-bladed propellers. One 5-AT was manufactured with its wing engines installed right in the wing, very much as in Stout's ill-fated 3-AT plane. But the speed gained was only negligible, and only one plane of that type was built.

Some 5-ATs had a larger passenger capacity than others. After more than forty 5-ATs had been manufactured, the fuselage of the model was extended some six inches, which increased the volume of the passenger cabin by approximately sixty cubic feet. This in turn increased passenger capacity to seventeen passengers—four more than the capacity of the first 5-ATs. Later, 5-ATs also had improved passenger accommodations. Over each seat individual electric lights were installed. Also, the planes had heating and ventilation systems as well as better lavatory facilities.

Some Ford Tri-Motors had different types of skin. Ford made available—at an extra charge—a special "Birdseye Finish," which some customers purchased. Consisting of a circular rubbed pattern put on the metal skin before it was cut and formed, this finish gave the plane a shinier and more scintillating appearance. At the opposite extreme, one 5-AT was built with a smooth metal skin instead of the familiar circular rubbed pattern put on the metal skin before it was cut and formed.

Some 5-ATs were modified to meet customer specifications. On some planes, baggage bins were built into the Ford Tri-Motors' wings. On others, such as a 5-AT purchased by Pan American-Grace Airways, hatches were cut into the fuselage so heavy machinery could be lowered inside for transportation by air.

Sometimes a new model evolved from the 5-AT. One example was the 6-AT—actually a 5-AT converted to a seaplane by replacing its wheels with king-size pontoons. Purchased by the Royal Canadian Air Force, this plane was used for crop dusting. No plane this size had ever been used to dust crops. But in experiments the 6-AT was able to carry one thousand four hundred pounds of arsenate dust and lay a 250-foot-wide path of pest extermination. Sometimes it flew less than twenty feet above the treetops. Other 5-ATs were modified with skis.

Then there were the 5-AT "Club" models. Advertised as "Air Yachts," they were luxury planes costing approximately $15,000 more than standard 5-ATs. With elegant, individually styled interiors, these planes came equipped with special furniture such as sleeping berths, divans, ice boxes, radios, typewriters, and desks.

One "Club" 5-AT was purchased by Marshal Chang Hsueh-Liang, a Chinese warlord. With soft upholstered swivel chairs, it was also decorated with a special wallpaper. Another "Club" 5-AT was bought by Prince

George Bibesco, president of the Roumanian Federation Aeronautica Internationale. Equipped with a kitchen and a main cabin that contained a bed, the plan was luxuriously fitted. Unfortunately, less than a month after its formal christening, as Bibesco and three companions were flying over India en route to Saigon, the plane collided with vultures in midair near Allahabad and crashed, seriously injuring Bibesco.

Despite the superiority and popularity of Ford Tri-Motors, orders at the Airplane Manufacturing Division fell off drastically in 1930 and 1931. In all of 1930, only 5-ATs were manufactured, and the following year, production numbered only nineteen planes. This of course was a direct result of the Great Depression. With Fokker now out of the picture, Ford Tri-Motors were just about the only passenger transport available. However, although the airlines had been expanding their routes and facilities under the stimulus of the Watres Act, they had little money left over in the rock-bottom economy to afford the large capital outlays needed to purchase new airplanes. Then, too, in the economy-minded era, the airlines saw the advantage of sticking with the fleets of Ford Tri-Motors they had already acquired and paid for. Did not the Airplane Manufacturing Division itself promote the fact that 4-ATs and 5-ATs had a usable life of at least four years? Well, most of the Ford Tri-Motors flying had several years more to go.

So, although the drastic drop in sales was upsetting to Henry and Edsel Ford, it was not necessarily a cause for concern. On the other hand, what was a cause for concern was the future of the Ford Tri-Motor. Once the effects of the depression lessened and the airlines started purchasing planes again, would the Ford Tri-Motor be the plane that met their requirements? Various indications were that the answer might be "No."

It was not only that the Ford planes themselves might become out of date, but also that a plane with three engines would no longer be the most desirable. The reason for this was that new improved engines had been developed which could give an airplane the same power with one or two engines that it previously had had with three engines.

Wright Aeronautical Corporation, for example, had recently introduced a new engine it called the Cyclone that could deliver 575 hp—almost three times the power of the original Wright J-4 Whirlwind engine used in the 4-AT. Similarly, Pratt & Whitney had introduced a new 575 hp engine called the "Hornet." With such powerful engines, two could do the job of three older engines, and they could do it less expensively, since they used less gasoline and oil and needed less overall maintenance. In an industry made cost-conscious by the depression, three engines could well become a luxury.

Sensing this trend, the Fords made a few tentative moves in the direction of single-engined planes. For several years, owners of 4-ATs that had been operated for long periods had been sending their planes back to Dearborn for reconditioning. Standard procedure had been to return the planes with new Wright Whirlwind engines similar to the ones with which the planes had been originally equipped.

But in 1931 the Airplane Manufacturing Division wrote all 4-AT owners of a new reconditioning program by which the three-engined plane would be returned a single-engined plane, equipped with one of the new Wright Cyclones or the Pratt & Whitney Hornet. The removal of the two outboard engines, the company explained, would result in a lighter plane (because the gasoline tanks for the two removed engines would be eliminated) and a more streamlined plane. The result would be an all-metal single-engined Ford plane with greater speed and bigger cruising range.

Although the Airplane Manufacturing Division stressed in its letter that the conversion program was recommended only for those 4-ATs used in freight operations (for passenger safety, three engines were essential, it said), rumors started that the new reconditioning program for 4-ATs was the first step of the Ford Motor Company toward introducing a new product line of single-engined passenger and freight planes.

Henry Ford himself denied this rumor, but when the National Air Show opened in Detroit on April 12, 1931, one of the three Ford planes on display was a "Ford Freighter," looking exactly like a Ford Tri-Motor except that it had only one engine.

The other two Fords on display were trimotors. One was a standard 5-AT. The other was a special 5-AT that had two Wright Whirlwind engines under the wing but a more powerful new Wright Cyclone engine in the nose. Except for these two trimotors and one Stinson Detroiter trimotor, every one of the other two hundred aircraft exhibited at the National Air Show was a single-engined plane. The Bellancas, the Northrops, the new Lockheed Orion—none had three engines.

The show, held at Detroit Municipal Field, was an exciting one, and more than fifty thousand people attended. Teenagers seemed to predominate, and even smaller children were there, dragging their parents from plane to plane, explaining to them such principles as the autorotation used in autogiros and the new electrical inertial starting systems for engines.

But the hit of the National Air Show was the single-engined Ford Freighter. When Henry Ford arrived at Detroit Municipal Field with one of his grandchildren, the press crowded around. Was Ford going to abandon the trimotor and concentrate on single-engine planes now? Ford parried all questions. The Ford Freighter was an experimental plane, and it was intended for transporting freight, not passengers, he said. For passengers, the three-engined 5-AT was still the best airplane around.

Later the Ford Freighter was sold to Pacific Alaskan Airways, a Pan American Airways subsidiary, and no other single-engine plane was ever produced by Ford.

Actually, the Freighter was one of a dozen different experimental models built by the Airplane Manufacturing Division at the same time it was manufacturing its regular 5-ATs. Often these experimental planes reflected Henry Ford's aeronautical whims. Shortly before, Ford had got the idea that diesel engines could be used in aircraft. This idea evolved from his dissatisfaction with the aircraft engines then available. "They are really automobile engines we just happen to use in aircraft," he wrote in a by-lined article in *American Magazine* in July, 1929. According to Ford, new thinking was needed on the subject of airplane engines. He urged young engineers to turn their attention to this subject, and he suggested as a possible lead that diesel engines might be easier to use and more reliable in aircraft than gasoline engines.

David A. Weiss

The reaction of the manufacturers then producing engines was characteristically negative. Charles L. Lawrance, the Wright Aeronautical Corporation president, admitted that the engines then available left much to be desired, but he said that of all the types of engines that should be explored, the diesel had the least potential. They were too heavy, he said, and needed too much strength for the power stroke. Rentschler, speaking for Pratt & Whitney, was equally against the idea of diesel engines for airplanes. He too said they were too heavy and ineffective.

Ford's answer was to build a diesel-engined Ford Tri-Motor. Three Packard diesel 225 hp engines were installed on a 4-AT owned by Ford Air Transport Service. The plane did fly satisfactorily, but the diesel engines were not that much of an improvement. Shortly after, the plane was converted back to a regular 4-AT. No other diesel-powered Ford Tri-Motor was ever built.

Another experimental 5-AT of which little is known was an attempt of the Airplane Manufacturing Division to build a Ford Tri-Motor bomber for the United States Army. A regular 5-AT was modified secretly for the purpose of entering it in an army design competition. Designated the XB-906, this plane was fitted with bomb racks, had a glass "greenhouse" in the bottom of its nose for bombardiers, and several open cockpits cut out of the fuselage for machine gunners.

The XB-906 was flown to Wright Field, Dayton, Ohio, for test checks. It evidently passed satisfactorily. However, soon after, back at the Ford factory, it had an unfortunate and premature end, crashing during more tests and killing its pilot, Leroy Manning.

The biggest Ford Motor Company aircraft experiment—and the company's largest and most costly failure—was a giant trimotor three times the size of any land plane built in the world up to that time. This was the plane whose existence Mayo had leaked to the press in San Diego and that Hicks had been working on all these years. Its once-publicized capacity of one hundred passengers had been scaled down, but when finally built, it still had a capacity of thirty-two passengers—almost twice as great as the capacity of the largest 5-AT flying.

Designated the 14-AT, the giant plane measured 110 feet between its wing tips, and its fuselage was more than 80 feet long. The 14-AT was also the world's most powerful plane, with a total of 2540 hp. Each of the outboard engines was a water-cooled 720 hp Hispano-Suiza. The third engine was a 1100 hp "W" type Hispano-Suiza. To achieve the ultimate in streamlining, the outboard engines were practically hidden inside the thick wing of the giant plane. The third engine was mounted in an ultrastreamlined nacelle over the midsection of the wing. It had a three-bladed steel-alloy propeller, while the outboard engines had wooden propellers.

Passenger accommodations in the 14-AT were far ahead of their time. In all, there were four passenger compartments, each of which seated eight passengers. At night the seats could be converted into sleeping berths exactly like Pullman berths. There were two washrooms on the plane, and also a large galley complete with stove and refrigerator. For soundproofing, the plane's walls had double construction. The dead-air space between provided a better insulation than any yet available on airplanes. Also as an innovation there was a thermostatically controlled heating and ventilating system. And as more or less proof of the "all-outness" that Henry Ford had ordered for this plane, it had a room for smoking—something Henry Ford abhorred. The pilots' compartment in the 14-AT was also unusually large. Big square glass windows all around offered visibility in every direction, and there was room enough for a flight deck.

Unfortunately, the 14-AT never flew. On the one attempt to get its tail into the air, the rear wheel of the giant plane rammed through the rear control surface. For reasons the Ford Company never revealed, no more attempts were made to get the 14-AT off the ground. But obviously the plane was as unflyable as Stout's 3-AT, years before. According to one story, Henry Ford was responsible for the failure of the 14-AT. Hicks had designed the giant plane to use the new Pratt & Whitney air-cooled engines, it was said, but at Ford's insistence, the heavy, untried, water-cooled Hispano-Suiza engines were installed. True or not, the plane was never shown to the public. A rumor started that it would be exhibited at the 1931 National Air Show, but it never was. As aviation historian Douglas J. Ingells summed up the 14-AT: "(It was) the plane of tomorrow, born yesterday, doomed to die before she spread her wings."

The failure of this giant plane produced an atmosphere of gloom over the Airplane Manufacturing Division. The morass of the Depression, the severe drop in sales, the increasing interference of Henry Ford in design, the apprehension over the future of three-engined planes—all were taking their toll on morale. One by one those who had contributed so greatly to the Ford Tri-Motor began to drift away. Mayo left soon after the 14-AT was completed. Hicks made ready to depart.

And Stout, who had hung on all these years, decided it was time to take his leave. Although he had had nothing to do with the giant plane, he waited until it had been built and then requested permission from Edsel to resign from Ford Motor Company and to start up his Stout Engineering Laboratories again. With a warm feeling for the man who had guided the Ford company into aviation, Edsel did not hesitate. He gave his approval.

To Henry and Edsel, who were left, the future was bleak at best. The airlines were still expanding and operating, but they were able to keep going only because of the subsidies they were receiving under the provisions of the Watres Act. So deep were the inroads of the Great

The giant 14-AT, Ford's largest and most costly airplane failure. It was three times the size of any land plane built in the world at that time, FORD ARCHIVES PHOTO

Depression that many a commercial passenger flight was flown without a single paying passenger.

In view of the drastic financial situation of the airlines, Ford Motor Company had few sales prospects for 5-ATs. The company had done everything possible to stimulate sales. It cut prices again. It kept adding improvements. But the sales statistics spoke for themselves. In 1932, only two Ford Tri-Motor s were sold.

As Stout had begun thinking about his Stout Air Services two years before, the Fords were coming to the conclusion that perhaps they had gone as far in aviation as they could. Although on paper the Airplane Manufacturing Division had shown profits some years ($4,437 in 1925 and $187,784 in 1929), the company had really lost money in aviation—since not taken into account in these figures were the thousands of dollars the Fords had spent on some of the experimental planes, the construction and maintenance of Ford Airport, and the operation of Ford Air Transport Service.

In 1932, when more accurate statistics became available on the Fords' investment in aviation, they indicated that the Airplane Manufacturing Division alone had cost six million dollars and that, when all the

expenditures involved in Ford Airport and Ford Transport Service were added, the total loss was $10,300,000.

The Fords could and did absorb these expenditures without much difficulty, but the question became: How much longer did they want to continue doing so? Even if sales did pick up, the company—with its production-line methods—could in six months produce enough planes to keep the airlines supplied with Ford Tri-Motors for another four years. That was assuming the airlines would still want Ford Tri-Motors. Many signs indicated otherwise. There were indications, for example, that airlines desired planes that were more comfortable than the Ford Tri-Motor. American Airlines, for one, had recently purchased some Curtiss Condors, huge biplanes that also offered convertible sleeping facilities, like those of railroad Pullmans. With a capacity of twelve passengers, the Condors were put on American's New York-Chicago run, and the airline advertised the Condor's advanced ventilation and insulation, areas in which the Ford Tri-Motor was deficient.

Yes, if anything, three-engined planes like the 5-AT were on the way out. Although no other types of planes were on the market yet, rumors had started that Boeing was working on a sleek-bodied two-engined plane that would be faster and more powerful than the Ford Tri-Motor. Another almost unknown company, Douglas Aircraft, was also thought to be developing a speedy two-engined plane.

In its blueprint files the Airplane Manufacturing Division had drawings of fast, new, two-engined planes just like the Boeings and Douglases were supposed to be. But with all the effort put into experimental models like the flivver plane and the giant 14-AT, these other planes had never gotten off the drafting table. Exhausted from their seven years of effort directed at trying to make commercial aviation a reality, the Fords began to question whether an additional effort was really worth it.

The Curtiss Condor. Although a biplane, it offered features the Ford Tri-Motor did not have, such as Pullman-type "sleeper" berths and advanced insulation and ventilation, EASTERN AIRLINES

Maintaining a large factory with sixteen hundred employees seemed like an extravagance when in 1932 only two planes were sold. Other factors, too, were turning the Fords against aviation. One was the loss Henry Ford still felt over Harry Brooks's tragic death. From the time of the crash of the flivver plane, something disappeared from Henry Ford's heart as far as aviation was concerned. Then there was the problem with Dr. Hugo Junkers. His German aircraft manufacturing company had been producing passenger planes with corrugated sides as far back as 1924. Moreover, some of these planes had been seen in the United States, and years before Dr. Junkers had visited Henry Ford at Dearborn. Although it seemed likely that Stout based his designs more on independent engineering thinking, Junker did not see it that way and began threatening Ford with lawsuits, claiming the corrugated skin and cantilever-wing construction of the Ford Tri-Motor had been directly copied from him. When the Fords refused to negotiate a settlement with him, he secured an injunction from a court in Spain, and seized there a Ford Tri-Motor that had just been delivered to Concesionaires de Lineas Aereas Subvencionadas S.A., the government airline. The Ford Motor Company managed to get the plane released, but for two years Junkers kept fighting in the courts. The Fords felt that Junkers's claims would eventually be turned down by the courts (which they were), but while he continued his legal action, any chance of Ford Tri-Motor sales in Europe was decreased tremendously. No foreign purchaser of a 5-AT wanted to take a chance that his airplane would be confiscated by a court to appease an angry litigant.

The Junkers Ju 52. Like the Ford Tri-Motor, this German plane had corrugated-metal construction and three engines. Also, like the Ford Tri-Motor, this 1930s plane had a reputation for dependability and reliability and even had a nickname—"Old Auntie Ju." LUFTHANSA GERMAN AIRLINES

Besides all these reasons that pointed up the desirability of getting out of aviation, Ford had some not even connected with his aircraft activity. The depression that had all but wiped out Ford Tri-Motor sales had also wreaked havoc with Ford automobile sales. Complicating this situation was the necessity to bring out a new automobile model (the V-8) to replace the Model A. In addition, Ford was having increasing problems with workers at the Ford Motor plants. Ford had once been a hero to labor. His establishment of wages of $5 per day when other companies were paying $1 and $2 daily had been praised by workers all over the nation. But throughout the following years he had become reactionary. In the face of layoffs resulting from the depression, some of his auto workers marched to the Ford plant at Dearborn to petition for more jobs and improved working conditions. Unfortunately, agitation developed, and the workers were fired on by police using machine guns and tear gas. More recently, the United Auto Workers of America had tried to organize the Ford plants. Ford bitterly opposed the union, but the National Labor Relations Board upheld the union's right to organize, and Ford became even more embittered.

With everything else, aviation seemed too much, and in July, 1932, those Airplane Manufacturing Division employees with automobile experience were transferred to the Ford company's automotive division. All others were discharged. The public was told nothing, and not for three months was the first report published that the Ford company had stopped making Ford Tri-Motors. Henry Ford confirmed the story, and said that if and when the company ever returned to manufacturing airplanes, it would produce single-engined and two-engined planes.

Rapidly all aviation activities at Ford Motor Company came to a halt. Within a few months, Ford Air Transport Service was discontinued, and its fleet of 5-ATs disposed of. Ford tried to sell the planes to an airline, but none could afford the capital investment involved in so many planes. The United States Navy finally came to the rescue, and acquired the planes at bargain prices.

A skeleton staff managed to complete the remaining unfilled orders. There were only three—one 5-AT for the National Air Transport division of United Airlines, and two for Pan American Airways. All three Ford Tri-Motors were completed in 1933. The last plane off the production line was one of the Pan American planes. On June 7, 1933, it was tested at Ford Airport and then turned over to the airline. It was the last Ford Tri-Motor ever manufactured.

Chapter Ten
New Passenger Planes Take Over The Routes

For seven years, almost to the day, the Ford Motor Company had been building 4-ATs and 5-ATS, a total of 199 Ford Tri-Motors in all, and if you count the special planes like the 6-AT sea plane and the Ford Freighter, which was really of trimotor design although it had only one engine, the count becomes 201 since that first 4-AT was flown successfully on June 11, 1926.

Considering the impact these planes had on American aviation, it was a small number. Yet, with this production and the Ford's many peripheral activities in aviation, Henry and Edsel Ford had accomplished all they had set out to do. From the time they had first become involved with William B. Stout and the Stout Metal Airplane Company, they had said over and over that all they wanted to do was to prove that commercial aviation was safe, reliable, and economically feasible. This they had done.

With the Ford Tri-Motor they had given American aviation its first practical and modern commercial airplane. With the Ford Air Transport Service they had demonstrated that a freight-carrying airline could be operated as smoothly as a railroad. With their support of Stout Air Services they had shown that passenger travel by air could be made attractive and comfortable. With Ford Airport they had pioneered the technical operation of the modern airport. With their sponsorship of the Reliability Tours and the advertising ads they ran telling the public that passenger air travel had arrived, they had educated the American public to accept aviation as reliable and safe.

As for their decision to back out of aviation when they did, it seems in retrospect their best course of action. As historian Allan Nevins put it, "Ford was concerned primarily with the manufacture of motor cars, and

Preceding Page: Loading mining machinery into the hatch of a Pan American-Grace Airways Ford Tri-Motor for flight over the rugged Andes, PAN AMERICAN WORLD AIRWAYS

no one can argue that he was foolish, in a time of stringency, to abandon an unprofitable side activity."

No big headlines screamed the news that no more Ford Tri-Motors would be manufactured. Even in the aviation industry, the future unavailability of these airplanes produced no particular excitement. The major airlines were well stocked with Ford Tri-Motors, and, when they were able to afford new planes, they would probably turn to the new Boeings and Douglases they had been hearing so much about.

At the time production of the Ford Tri-Motors had ceased, neither Boeing nor Douglas had introduced their new planes, but they had been developing them, and their introduction was not far off.

As far back as 1930, Boeing had brought out a single-engined cabin plane called the Monomail. Built around the Pratt & Whitney Wasp engine, this plane was the forerunner of the sleek modern planes to come, but, with only a single engine, it was underpowered. Boeing then moved on to its famous Boeing 247 plane, which had twin engines. The intention had been to design the 247 around the Hornet, the newest and most powerful Pratt & Whitney engine, but pilots protested. Accustomed to the lower-powered Wasp engine, they insisted this be used. In acceding to this demand, Boeing had to reduce the number of passengers it intended the plane should carry, but it was still able to obtain a maximum speed of 200 mph and a cruising speed of 160 roph. No plane this fast had ever been seen before in American commercial aviation.

Since Boeing was locked in corporately with United Airlines, its sister company, United tied up all initial orders for the new plane. TWA, anticipating this situation, had already made other arrangements. Its engineers had worked up specifications for a new commercial plane, and it asked various airplane manufacturers other than Boeing to bid on producing it. One bidder was Douglas Aircraft, which up to that time had specialized in turning out military and mail planes. Douglas Aircraft was operating on a shoestring in a small Santa Monica plant, but it was headed by a young engineering genius named Donald Douglas. Jack Frye, cofounder of Standard Airways and now vice-president of operations for

TWA, recognized Douglas's potential. Frye called Douglas in for further consultation, and, as an outcome of these meetings, Douglas Aircraft was awarded the contract to produce TWA's new plane.

The Boeing 247. Able to cruise at 171 mph with ten passengers and a crew of three as well as cargo, this plane could fly coast to coast in less than nineteen hours, UNITED AIRLINES

The result was the DC-1, an all-metal, twin-engine plane that could carry twelve passengers at speeds faster than could the Ford Tri-Motor. When Douglas completed the first DC-1 for TWA, Frye and World War I flying ace Eddie Rickenbacker went to Santa Monica to pick it up. Their flight in the new plane from the West Coast to the East Coast established a record—they made the trip in thirteen hours and four minutes.

No sooner was the DC-1 in production than Douglas came out with a larger plane, the DC-2, which could carry two more passengers than the DC-1. Now working with American Airlines as well as TWA, Douglas realized that an even larger plane was needed to make the DC model economically attractive to the airlines. The result was the DC-3—probably the most famous plane in the history of American commercial aviation.

The DC-3 had everything. With a maximum speed of 212 mph and a cruising speed of 180 mph, it was faster than the Boeing 247. Boeing had been restricted in its engines to using those of Pratt & Whitney, its fellow United company. Douglas was able to use new Wright Cyclones, each capable of developing 1,100 hp. The DC-3 could also carry more passengers (it had twenty-one seats) and a bigger payload (4,900 pounds). It also had many safety features, among them, automatic pilot, wing flaps, and the ability to fly on one engine. It had many advantages for passengers, too, such as steam heat in the cabins and soundproof walls. So superior was the DC-3 that by the late 1930s it was carrying almost 90 percent of the world's air commerce. More than twelve thousand were manufactured in all—sixty times the number of Ford Tri-Motors.

The introduction of the Boeing 247 and the Douglas DC-3 in 1933 and 1934 sounded the knell for the Ford Tri-Motors still flying on major United States airline routes. Just as the wooden Fokkers several years before had vanished almost overnight, so now did the Fords. The old planes with the corrugated-metal skin were simply too slow, too noisy, and too lacking in personal comfort.

The DC-3. By the late 1930s these planes were carrying almost 90 percent of the world's air commerce. AMERICAN AIRLINES

As one by one the Fords started moving off the major air routes, the pilots who flew them waxed nostalgic. Said one TWA pilot: "Like faithful old fire horses, unshod and retired to a life of ease and green pastures, a fleet of veteran trimotors is facing retirement after almost a decade of active service on the mid-transcontinental airway between New York and Los Angeles."

Said John Collings, a onetime Ford test pilot, later a TWA executive: "The end is in sight for the old girl. She was and is a thoroughbred, but her days are almost up."

It was at this time that the name of "Tin Goose" began to be applied to the Ford Tri-Motor; although no one now knows when, where, and by whom it was first used. Larry Fritz, one of the early test pilots, said he heard it around the Ford plants as early as 1927. Another pilot, D. W. Tommy Tomlinson, used the phrase in an aviation article published in 1929, and a writer named Ralph Hancock claimed to have invented the term.

But the affectionate term did make sense. When the planes were new and passenger flying still something for a very few, the 4-ATs and 5-ATs were formally called "Ford All-Metal Planes." When they were first put on major routes, they dwarfed all other commercial planes, and when plowing through the air they did truly seem majestic. "Like giant condors they were in the air," a contemporary observer noted, "like queens of the sky."

But on the ground, even in their heyday, their bouncing along the runway on three wheels for a takeoff bore a striking resemblance to a goose waddling down a road in front of her goslings, and the assorted noises that the plane made as it flew sounded vaguely like the honk of a goose. The Ford Model T was affectionately called the Tin Lizzie, wasn't it? So why not the Tin Goose for this other famous Ford product?

It might seem to a foreigner that these phrases were ones of derision. Even those who spoke them seemed to be a little contemptuous of the tin box on wheels and of that noisy, rattling airplane that waddled down the runway. But down deep, people felt admiration and pride. Damn! These Ford products were sturdy and reliable. They might look tinny, but they were strong, they never wore out, and, damn again, they kept going forever.

Not even Henry or Edsel Ford thought the Ford Tri-Motors had an operational life of much more than four years. In advertising the planes in 1929 they proudly stated that no Ford yet had worn out in service. "Consequently, Ford planes run little if any danger of becoming obsolete before they complete their period of usefulness—a period we know now to be not less than four years."

By 1934, when the last of the Ford Tri-Motors were pulled off the major air routes, many of the planes were nearly six years old. Many, too, had been greatly overworked. But operationally—despite the Ford prediction of four years—they were as good as ever.

For that reason they were not retired, but were put to work on the less glamorous shunt routes. Sometimes they flew these routes for the same airlines that had originally purchased them, other times for smaller airlines to which they had been sold. In 1934, when Braniff Airways acquired a small Texas line called the Long & Harmon Air Service, it also acquired two Ford Tri-Motors, comprising Long & Harmon's complete air fleet. For two more years Braniff flew these trimotors over Long & Harmon's air routes, until they finally were replaced with Lockheed Electras. Even that did not mean the end of the Tin Geese. Braniff sold them to a smaller airline, and they kept flying.

And so the story went for practically all the Ford Tri-Motors remaining of the 199 originally manufactured by Ford Motor Company. Replaced they were, on route after route, but seldom were they grounded. Patched up, rehabilitated with new engines, re-registered with other airlines, often in other countries, for someone—somewhere—they kept on flying and it wasn't long that some got a new lease of life south of the border. in parts of Mexico, Central America, and South America where the topography made railroad and highway construction virtually impossible. High mountains, swamps, broad rivers, barren deserts, impassable jungles—hundreds of villages, towns, and mines were effectively isolated from the rest of the world. Boeings and Douglases could not be flown into such areas. Their speed and power required airfields with long runways, and these could not be built in such rugged terrain. But a Ford Tri-Motor that could lift itself into the air, even with a heavy payload, after traveling across the ground for less than one thousand feet—a plane like this was ideally suited for bush operations. Tin Geese could fly in and out easily, and at costs far less expensive than mule train or native labor.

One Central American airline purchased as many as twenty-seven secondhand Ford Tri-Motors for this type of bush operation. This company was Transportes Aereos Centro-Americans, better known as TACA. One of the largest carriers of freight and passengers in the northern hemisphere, TACA was founded by an ex-World War I pilot named Lowell Yerex. Barnstorming through Central America in the late 1930s, Yerex was stranded in Mexico City. There he was approached by three American mining engineers.

"We're going up to Yucatan to mine gold," they said. "We have a plane, but no pilot. Want a job?"

A Panagra Ford Tri-Motor gets ready to load passengers at the airport in Pacasmayo, Peru, in 1930, PAN AMERICAN WORLD AIRWAYS

The cabin of one of National Air Transports (NAT) Trimotors flying the Kansas—Dallas route. NAT was the forerunner of both United Air Lines and TWA. UNITED AIR LINES

Within six hours Yerex was flying it north to Yucatan. Within six months, the plane was his. When the mine did not pan out, the engineers were unable to pay Yerex his wages. They gave him the plane.

With this one plane, Yerex started TACA. Headquartered in Tegucigalpa, Honduras, the airline soon added branches in Costa Rica and Nicaragua. It also started buying Ford Tri-Motors. "The best plane ever built," said Yerex. "I could have used a hundred of them."

TACA pilots never knew where or what they would be asked to fly next. Sometimes it was mining equipment; other times, ore; still other times, fuel or food supplies. Often the Tin Geese would have their fuselages modified. One old 5-AT was converted into a flying tanker that could carry six hundred gallons of diesel oil. In one day this plane made four trips into the mountains of Honduras, bringing twenty-four hundred gallons of desperately needed oil to a silver-mining camp.

For more than twenty years TACA flew Ford Tri-Motors into the jungles and mountains of Central America. One of its typical operations was to bring chicle out of a plantation in Yucatan. With the sun blazing overhead, natives would pile the chicle by the side of a tiny jungle clearing. Suddenly the engines of an airplane would be heard. Out of the sky would come an ancient battered Tin Goose. Heading for the small clearing, it would come down, landing with a bounce, its wheels kicking up clouds of dust. Grinning, a pilot would jump out and talk to the plantation foreman as the natives loaded the chicle into the hold. Then he would jump into the plane again and take off quickly, the plane soon a dot over the mountains. It was a far cry from the swank Transcontinental Air Transport coast-to-coast service the Ford Tri-Motor had once flown, but the plane was still doing a job.

Small fleets of Ford Tri-Motors were also used in other parts of Latin America. Four old corrugated-metal Tin Geese shuttled food and mining supplies over the Andes to isolated mining camps. Once at Huanacopampa, Peru, Ford Tri-Motors flew in 740 tons of mining machinery from Cusco. On each flight the planes had to climb more than fifteen thousand feet to clear the high mountains.

Ford Tri-Motors were also used for oil exploration in the upper reaches of the Amazon River. One of these operations was directed by Rex Noville, once an executive officer on a Byrd South Pole expedition. As late as the 1950s, Noville was still using Tin Geese.

Ford Tri-Motors found their way to all parts of the world. In far-off Australia, the Royal Australian Air Force converted a 5-AT into an ambulance plane. In China, explorer Dr. Joseph Rock flew from province to province in a chartered 5-AT. Once he flew from Shanghai to Yunnan in 42 hours. The same trip several years before by land had taken him fifty days. From the matted jungles of New Guinea to the icy waters of Alaska, Ford Tri-Motors kept flying . . . and flying . . . and flying.

Many were still flying in the United States too. Some of the planes purchased by corporations never left American shores. If the corporation no longer had need of them, they often sold them to other corporations. The 4-AT purchased by Curtis Publishing Company was later sold to

Arthur Kudner, the well-known advertising executive who coined the phrase "athlete's foot" for Absorbine, Jr. For years Kudner flew his Ford Tri-Motor in making his rounds to his advertising clients.

And then, just before World War II, Ford Tri-Motors had a revival in the United States. They became a barnstorming plane, much as the Curtiss Jennys and De Havilland DH-4s had in the early 1920s. At county fairs and carnivals all over the nation the Ford Tri-Motors would show up, landing at local fairgrounds or in nearby cow pastures, offering to take up any and all for a ten-minute ride at a dollar a head.

Sometimes, the big old Tin Geese would make their local sightseeing flights at night. Sitting on the improvised landing field, the plane would be lit up like a carnival ride, with strings of electric lights tied to its wing and fuselage. Kids holding their parents' hands would stand in awe, looking at the big battered plane and listening to its three engines roaring. Then, fifteen at a time, the passengers would board for the ten-minute flight. With lights still blazing, the plane would take off and circle over the nearby town. Once directly over the town square, the pilot would let loose a string of Roman candles. Instantly the sky would be filled with dazzling arrays of fiery streamers and multicolored fireballs.

The fireworks were the crowning touch, but the memory that would linger with the kids was of the flight itself, their first flight. "I flew, Mom, I flew," they would say excitedly after they landed. And for hundreds of thousands of Americans those flights in the barnstorming Tin Geese were their introduction to aviation.

Ford Tri-Motors also flew in the spectacular air circuses held in conjunction with the air races that took place every year in the United States. Fliers like Jimmy Doolittle and Roscoe Turner flew for trophies like the Bendix, Harmon, and Thompson. Accompanying these events were aerial circuses featuring aerobatics, comedy routines, clowns, and pretty girls.

The barnstormers now traveling with the air circuses were no longer unshaven loners drifting from town to town in dilapidated Jennys. Now they were heroes as popular as the racing pilots, and they flew wonderful shiny planes with unbelievable strength and with engines that screamed power. One of these new barnstorming heroes was Beverly E. "Bevo" Howard, considered the greatest precision aerobatics pilot ever to fly in the United States. He flew a special Buecker-Jungmeister, the plane used to train Luftwaffe pilots. Another was Bob Hoover, who flew a North American P-51 Mustang. Another was Paul Mantz, who once flew forty-eight consecutive loops and could fly in a way, as a fellow flier once said, that would make an angel green with envy.

And then, considered by some the mightiest of them all, there was Harold Johnson, "King of the Tin Goose." Yes, Tin Goose! Those other aerial stuntists flew the most modern, powerful planes, and in models built just for them. But Harold Johnson stunted in one of the original Ford Tri-Motors that had not been altered or modified in any way. With this stock Tin Goose Johnson ran the gamut of unbelievable aerobatics. There was nothing he did not, or could not, do—from vicious hammerhead stalls to Cuban eights and whirling spins that started as low as one thousand feet.

One time in Detroit, after going through his repertoire of unbelievable stunts, Johnson was approached by Henry Ford himself. Ford had seen the show, and was impressed beyond belief. He talked a few minutes to Johnson, and then, before leaving, handed him an envelope. Inside was a check for ten thousand dollars.

Johnson's most spectacular Tin Goose stunting took place at the 1939 Cleveland Air Races. As one hundred thousand pairs of eyes watched, Johnson did a loop right in front of the grandstand. Then he circled the field and came in looping again, only this time his plane was upside down. So low did he fly, the tip of his rudder almost scraped the concrete runway.

Then Johnson pulled the Tin Goose into an abrupt climbing turn, and from here did snap rolls, slow rolls, Immelmanns, and other aerobatics usually attempted only by fighter planes. His climax was a special loop-the-loop. Coming across the landing field at three hundred and fifty feet, Johnson did two consecutive snap rolls, which are difficult in any plane. Coming out of the second roll only twenty-five feet off the ground, he then ripped upward into a wide, clawing loop, but this was no ordinary loop, because after one loop, Johnson looped again, and then again. Three loops in a row he made in the big old Tin Goose without a break, and after the last one, he snapped into a side slip that ended up in a perfect landing.

The crowd was on its feet, cheering and applauding, but that was the last performance of the King of the Tin Goose.

The following year the National Air Races were canceled because of the outbreak of war in Europe. Once again barnstorming and aerial circuses were grounded.

As United States involvement in World War II threatened, Henry Ford was contacted by various federal government departments to join the effort to prepare the nation for hostilities. In most cases, automobiles, trucks, and tanks were talked about, but so also were airplanes. Before Pearl Harbor, the Air Corps talked to Ford about producing pursuit planes. The President had initiated a program to produce fifty thousand planes, and Ford Motor Company was expected to be one of the major suppliers. Although Ford was an ardent pacifist, he agreed to do what he could on grounds of patriotism. But nothing further happened with the program.

Shortly after, Ford was asked to build airplane engines, and again he agreed to do what he could. "If it becomes necessary, the Ford Motor Company could with the counsel of men like Rickenbacker and Lindbergh under our own supervision and without meddling by government agencies swing into production of a thousand motors a day." But again nothing happened.

After Pearl Harbor, however, the story changed. In the giant Willow Run plant built and operated by the Ford Motor Company, the B-24 Liberator bomber was produced. While these Ford-produced bombers played an important role in the final Allied victory, the old Ford Tri-Motors did their share too in winning the war, although on a much smaller scale. In China, before the war even started, they transported supplies over the Hump for the Flying Tigers. In the Philippines, they performed valiantly, helping to evacuate army personnel and civilians from Bataan in the early days of the war.

At the height of the evacuation, the exhausted members of an engineering topographic battalion scanned the skies desperately over

Bataan. Pinned down by advancing Japanese troops, they had been waiting for the airplane promised days before to carry them and their valuable topographic plates to Corregidor. Suddenly the roar of airplane engines was heard overhead. "It's an Army Tin Goose!" one engineer shouted. "We're going to get out of here at last!"

The beat-up old Ford Tri-Motor landed nearby on a tiny clearing.

"The plane's overloaded," the copilot yelled, after he had helped the men pile in and load their plates in the dusty hold.

The pilot laughed. "Jump in. This old plane's been overloaded every trip so far."

Back to Corregidor the Tin Goose flew, landing the men on the crowded island fortress. Then it took off again for another part of Bataan. For weeks the battered old plane evacuated personnel and supplies, and then suddenly it was seen no more. One story had the plane so overloaded that it had fallen into the sea. Another said its wing had been shot off by Japanese artillery. Whatever the case, for the brief few weeks it had been able to fly, it had done a hero's job.

All over the Pacific in the early days of the war Ford Tri-Motors saw service. In the South Pacific, General George C. Kenney, General MacArthur's air chief, used them for transports for his Fifth Air Force until more modern planes could be obtained. In Australia, the Royal Australian Air Force flew three Ford Tri-Motors against the Japanese. All the planes were destroyed within the first year. And thousands of miles away, at Dunkirk, another Ford Tri-Motor helped evacuate some of the embattled English troops, getting them across the Channel to safety.

Toward the end of World War II Ford Tri-Motors turned up in odd places. One American diplomat spotted a battered old Tin Goose flying around Stalingrad. In Italy, after the surrender, some Ford Tri-Motors were flown by Italian Air Force pilots to drop ammunition and food supplies to the underground in Yugoslavia. Operating from a field near Torino, Italy, the planes flew by night, mission after mission, until the assignment ended.

Four years of operating! For some Ford Tri-Motors, the saga was only beginning.

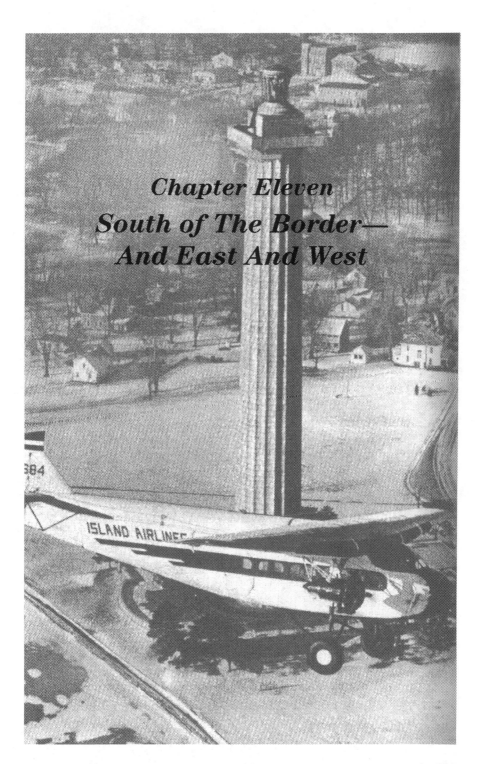

Chapter Eleven
South of The Border—
And East And West

From the 1940s on the Ford Tri-Motors kept flying. But as one year followed another, the ranks of the surviving Ford Tri-Motors slowly thinned out. In the 1950s, of the 199 originally manufactured, not more than thirty Tin Geese were still flying. By the 1970s, this total has dwindled to less than twenty.

Strangely, the last regular airline in the United States to use Ford Tri-Motors for passenger travel was headquartered not more than fifty miles from where the planes were originally manufactured by Ford Motor Co. in Dearborn, Miichigan. This was Island Airlines in Port Clinton, Ohio which provided an air-ferry service to several small offshore Lake Erie islands about halfway between Toledo and Cleveland. Advertising itself as "The Shortest Airline in the World," Island Airlines' entire route was only seventeen miles long.

The company traced its beginnings back almost as far as those of the Ford Tri-Motors it flew. As Island Airways, Incorporated, it was established in 1929 by an ex-barnstorming pilot named Milton "Red" Hershberger. With a fleet of Waco biplanes and an old Standard biplane, Hershberger at first carried only freight to the islands, but after 1935, when he began purchasing Ford Tri-Motors, he started carrying passengers as well. Through the years he sold some of the trimotors, but in 1953, when Island Airways was acquired by Ralph Dietrick, there were still two left. Dietrick, a World War II flight instructor and air-transport-command pilot, had been operating a similar air service from Sandusky, Ohio, to several other islands not on the route of Island Airways. Upon taking over, he sold his previous operation, moving the company's headquarters to nearby Port Clinton, Ohio, and changing the company's name to Island Airlines.

Opposite: An Island Airlines Ford Tri-Motor circles the Perry Monument at Put-in-Bay, on South Bass Island, in Lake Erie, ISLAND AIRLINES PHOTO BY RON KUNTZ

Although Island Airlines was an aviation jack-of-all-trades company, operating a flying school, air taxi service, and charter service as well as an airline, the mainstay of its business was its regularly scheduled flights. Except for Sundays, when it did not fly, the airline operated an average of five flights a day. On Saturdays it had nine flights, and, during the ice-fishing and hunting season, it put on extra sections when needed.

The Island Airlines service to the offshore islands was the granddaddy of the shuttle service that later started up between Washington and New York, and San Diego-Los Angeles-San Francisco; but its route was shorter and it made many more stops. Flying at about 85 mph in the old Tin Geese, Island Airlines pilots took less than forty-five minutes to cover the entire route, and this included twelve takeoffs and landings each trip. Yet, the payloads transported by the planes reached high totals as many as than 65,000 passengers a year in addition to 300,000 pounds of freight and 150,000 pounds of mail.

Passengers boarding one of the Island Airlines Ford Tri-Motor flights. "The Shortest Airline in the World," it flies 365 days a year, providing commuter service to and from the mainland and several offshore Lake Erie islands, ISLAND AIRLINES PHOTO BY RON KUNTZ

By the 1970s Island Airlines was operating three Ford Tri-Motors, two 4-ATs and one 5-AT. From the airline's base at the Port Clinton Municipal Airport, one of the Tin Geese would take off or land every ten minutes. Starting out on its route from Port Clinton on the mainland, its the first stop—eight minutes away—would be South Bass Island at the landing field at Put-In-Bay, so called because Commodore Oliver Hazard Perry "put in" here after his historic naval victory over the British fleet in the War of 1812.

The next stop, less than three miles away, was Middle Bass Island, where the George Lonz Winery was located. After Middle Bass came Rattlesnake Island, less than eighty-five acres in size with only three families as permanent inhabitants. Rattlesnake (named for its shape rather than its reptiles) had its own private post office, and sold there, as well as at the Island Airlines office at the Port Clinton Municipal Airport, special Rattlesnake Island triangular stamps, one of which featured a picture of the Ford Tri-Motor. The final stop, five minutes later, was North Bass Island, known for its grape harvests.

For the eight hundred inhabitants of these five islands, the three Ford Tri-Motors of Island Airlines provided a vital transportation link. In the summer, ferry boats also serviced the islands, bringing across tourists as well as automobiles which could not be transported to the islands any other way. But most inhabitants as well as the tourists prefered the air route, since the fare was not much more ($4.00 by air round trip to South Bass as compared to $2.50 by ferry), and the time was five minutes compared to one hour. In the winter, when Lake Erie frozes over, Island Airlines offered the only transportation.

So, here, off the shore of Lake Erie, the "mass" transportation was neither subway nor bus nor taxi, but Ford Tri-Motors manufactured in the late 1920s and the early 1930s. Those mainlanders who worked at the George Lonz Winery on Middle Bass Island commuted each day via these ancient planes. The islanders who wanted to shop for groceries in Port Clinton traveled back and forth by air. Ministers and physicians flew the Ford Tri-Motors in making their rounds on the islands.

Then there were the students who went to school by air. Although there was a small elementary school on Middle Bass Island, the nearest high school was at Put-in-Bay on South Bass Island.

No reservation was needed to fly on Island Airlines planes. All a passenger has to do is show up ten minutes before one of the scheduled flights. At Port Clinton, tags colored blue, red, or yellow were handed out. It was first come, first served. Only enough blue tags are given to fill the next plane. When the blue flight is run, it then became the turn of the yellow flight. With the tags passengers also received a small yellow ticket imprinted with a picture of a Ford Tri-Motor that could be kept as a souvenir, On it was a message: "Souvenir of your flight on The Shortest Airline in the World aboard a Ford Tri-Motor."

In some cases the first and last rides of the islanders took place in Ford Tri-Motors. Babies born in the Port Clinton hospital were flown back home by air and when someone on the island died, the deceased and his entire funeral party were flown back for burial on the mainland.

Besides passengers, freight of all kinds was carried by Island Airlines in its Ford Tri-Motors. Pets, seed, barbed wire, automobile tires, machinery, and medicine. Sometimes considerable ingenuity was necessary to carry a particular freight item. Once an island boatyard needed a special piece of lumber twenty feet long. To deliver it, Island Airlines had to position it in a Tin Goose so that one end stuck out the cockpit, and the other end, out an open door in the rear.

With the exception of Island Airlines and the Tin Gooses continuing to fly passengers and freight in Central and South America the days in the United States for the planes to be used in regular passenger air travel had ended after World War II. But toward mid-20th century a dozen new careers opened up for the Tri-Motors that in some cases had seen service for as much as a half century. Barnstorming, sightseeing, crop dusting, firefighting, touring, air shows, participation in anniversary celebrations of major airlines, even props in Hollywood films—the list goes on and on for the different uses to which they were put.

To fill these various functions the Tin Gooses were shifted back and forth among many different owners in what resembled a game of musical chairs, planes were bought and sold, leased and donated among different types of owners ranging from aviation museums, wealthy collectors of airplanes, and smaller airlines. So complicated did all this become that after a plane had undergone a succession of owners it sometimes ended up being bought by an airine or individual who had owned it years before.

No longer carrying passengers, many of the Tin Gooses ended up doing crop dusting or fire fighting like Johnsoin Flying Service, Missoula, Montana, which worked almost exclusively for the U. S. Forestry Service. At one time this company had as many as five Tin Gooses, some seeding grass or spraying oil emulsions to kill insects in soil redemption projects, other times fighting fires. carrying "Smoke Jumpers" to the scenes of fires (the Fords could hold eight men and their gear)), and dumping heavy borax-water mixtures in one large load from tanks with trap doors ahead of the fire to retard its spread.

In the fall of 1967, when one of the worst timber fires in thr nation's hstory was a blaze in a forest region of Idaho, Johnson Flying Service was first on the scene to parachute in men and supplies. Said Robert R. Johnson, owner of the company; "For this kind of flying, there never was and probably never will be a better aircraft built than a Ford Tri-Motor."

Another activity in which Ford Tri-Monitors werew widely used was in barnstorming. From town to town, city to city, a barnstomer would fly in, landing on everything from a pasture to a municipal airfield, showing off his Tri-Motor and taking those who wanted to fly in it on a short spin around the town., It was through these flights that hundreds of thousands of Americans who had never flown before flew for the first time. Neil Amstrong, the astronaut who became the first man on the moon, often mentioned in interviews that his first flight in an airplane was in one of these barnstorming Tin Gooses that that visited his home town when he was a small boy.

One of the many barnstormers in the early days was Al Chaney who painted his Ford Tri-Motor white red and blue, and flew around the Midwest Another was Charles LeMaster who with his wife took their Tin Goose, the "Kansas Clipper" around the Midwest. Then there was Al Chaney who for a half dozen years flew his Ford Tri-Motor up and down the East Coast and the Midwest in the winter months, and in the summer months in the southeastern states, particularly Florida.

Another barnstomer was was John Louck, operator of an airport in Monmouth, Illinois. A retired Air Force captain who had been a World War II flight instructor, Louck started his barnstorming operation after spotting a Ford Tri-Motor in Florida. It was in poor condition due to heavy wear, but after examining it carefully, he felt its wings and fuselage could be reconditioned, and he could make it serviceable. Louck put a lot of work into it, refurbishing the interior and installing new Pratt & Whitney engines. Once completed, the plane was sent barnstorming through the South and Midwest, its sponsor being Ford Tri-Motor Air Tours, a company he started to exhibit the plane.

Taking it to airshows and aviation meets, Louck found he could get all the passengers he wanted at three dollars a head. He took up about fifteen thousand passengers a year, some of them, ironically, pilots who had always heard of the famous old Tin Goose but who had never flown in one. At one air meet in Illinois in 1962, Louck had almost three thousand passengers, some of whom waited in line four or five hours to board the plane.

Meanwhile, some owners of Tin Gooses found themselves with a source of income they could hardly have anticipated: revenue from renting out their planes to Hollywood movie studios. It started in 1936 when a Ford Tri-Motor appeared in a 1936 Flash Gordon serial starring Buster Crabbe and continued into the 21st century with Public Enemies, the John Dillinger bio starring Johnny Depp released in 2009. Other films in which a Tin Goose appeared: The Fortune, Indiana Jones and Temple of Doom, The Untouchables, Raiders of theLost Ark, The Family Jewels, To Be Or Not To Be, and Only Angels Have Wings,

After their days of flyng passengers Ford Trimotors were often in demand to participate in anniversary celebrations from some of the airlines that had used them in their earliest days.

Through most of the second half of the 20th century Ford Tri-Motors were employed by the companies that did crop dusting and fire fighting in the Westand Mid-West. One important customer was Johnson Flying Service, Missoula,. Montana, which for years hadworked almost exclusively for the U.S. Forestry Service, and at one time had as many as five Tin Gooses. In soil conservation work, they were used to seed grass, and also to spray fine oil emulsions over fields to kill insects. In fighting fires they proved invaluable by carrying "Smoke Jumpers" with their gear to the scene of the fire and dumping heavy borax-water mixtures in one large load from tanks with trap doors ahead of the fire to retard its spread.

In the fall of 1967, one of the worst timber fires in the country's history hit the Sun Dance Forest region of Idaho. First on the scene to parachute in men and supplies was a Ford Tri-Motor of the Johnson Flying Service. Said Robert R. Johnson, owner of the company; "For this kind of flying, there never was and probably never will be a better aircraft built."

When eventually Johnson Fyimg Service began selling off its Ford Tri-Motors, the last one it sold was a 5-AT to Wings and Wheels, Inc. of Santee, South Carolina. Operated by Dolph Overton, a Korean War flying ace,Wings and Wheels maintained a large antique airplane museum on highways 15 and 301. In 1969, after arrangements were made to purchase the plane from the Johnson Flying Service, Jack Hughes, chief pilot for Johnson's, flew the old trimotor twenty-one hundred miles to South Carolina without a hitch.

This Ford Tri-Motor, along with another Ford Tri-Motor purchased previously by Wings and Wheels, was housed in a huge museum building that contained more than forty other vintage airplanes. Upon obtaining the Ford Tri-Motor from the Johnson Flying Service, Overton refurbished it and, besides displaying it in the museum, also used it to take people on sightseeing flights over the Santee-Cooper Reservoir.

Another museum with a Ford Tri-Motor at this time was Harrah's Auto Collection, Reno, Nevada. This plane was the eighth 5-AT built. First flown in December, 1928, it was delivered to National Air Transport the following year, and sold a few months later to Transcontinental Air Transport. TAT flew it on its coast-to-coast service for about two years, and then it ended up being used as a sightseeing plane by Grand Canyon Airlines. After 1937 the plane's whereabouts were unknown for a while, but during World War II it turned up first in Nicaragua and later in Mexico. In 1951, the Tin Goose was completely overhauled by Torres Landa in the shops of Servicios Aeronauticos de Mexico S.A., and emerged with a new smooth duralumin skin replacing the original corrugated-metal skin. Purchased next by Eugene Frank, who operated a crop-dusting outfit in Boise, Idaho, it was located there by Harrah's Auto Collection, and restored, winding up almost totally rebuilt, 98% new including a new skin. As such, it was described as the "most magnificent Ford Tri-Motor in the world."

Still another Ford Tri-Motor that participated in nostalgia was owned by Mox-Air. a barnstorming and exhibition service headquartered at Clover Field, Santa Monica, California. Devoted to collecting and restoring vintage airplanes, Mox-Air was owned by Gaylord Moxon, president of Moxon Electronics, which was also located in Santa Monica. Mox-Air through the years had built up a good collection of planes, including a Ryan, a Lockheed Vega, a Fairchild, a Curtiss Robin, and a Standard biplane, but it never had been able to procure a Ford Tri-Motor. Then, one day in 1964, Gaylord Moxon and his wife Grace were flying in their private plane over Idaho when suddenly Moxon spotted in a field below what looked like a Tin Goose. Landing at the nearest cow-pasture airport, Moxon and his wife rented a car and drove out to where they had seen the plane. It was a Tin Goose, all right, and it looked operational, but it was all beaten up. Used for crop-dusting, its inside cabin was stripped down, and tubes stuck out of the fuselage at various points.

The plane had been one of those 5-AT Tri-Motors that had flown on TWA's 1959 tour, and after that, had been had been converted into a plush executive plane, and from there had been taken to the Arctic where it flew over the Klondike for Alaskan Airlines. After that it was purchased by an owner who returned it to the United States where he used for a while in crop dusting and then in barnstorming country fairs and air shows.

Moxon bought the Tin Goose and after he got back to Santa Monica had it completely restored as it was on TAT's original 1929 coast-to-coast service. The same type of wood paneling was added, as well as the same seats, cockpit, instruments, and even engines. Original parts were used whenever possible, and more than six thousand man-hours of labor went into the restoration before it was completed.

Naming it Graceful after his wife, Moxon was soon using the rehabilitated Ford Tri-Motor for short sight-seeing flights over Santa Monica. Taking up fifteen passengers at a time ($6 for adults, $3 for children), Mox-Air also handed ut souvenir tickets that had on them a picture of the airplane and the words, "I flew in the original Ford Tri-Motor N-9651."

In September, 1967, when the port of Oakland, California, celebrated its fortieth anniversary, Moxon had the plane flown to Oakland International Airport to participate in the event. All types of modern aircraft were present, from military fighters and bombers to helicopters and jet-propelled airliners and spacecraft. But the hit of the show was Moxon's restored Ford Tri-Motor. "The ship is a living legend," Moxon said. "She has a real magnetism for drawing crowds."

David A. Weiss

The first airline to use the Ford Tri-Motor for promotional purposes was TWA. In 1949, the airline commemorated the twentieth anniversary of the inauguration of transcontinental service by reflying this trimotor over the original part-air, part-rail route. This is the original TWA Ford with its pilot, Art Bums, TRANS WORLD AIRLINES

Not unexpectedly, some of the remaining Ford Tri-Motors at this time were owned by major airlines which used them for promotional purposes. The first such airline to think of this was TWA which in 1949 decided to commemerate the twentieth anniversary of the inauguratiom of transcontinental air service by its original company Transcontinental Air Transport. Leasing a Tin Goose from Island Airlines, it repainted the plane with TAT markings and flew it over the part-rail-part-air route over which it had pioneered coast-to-coast air service.

Another major airline that used a Ford Tri-Motor to celebrate its anniversary was Northwest Airways (now Northwest Orient) which in 1969 celebrated its thirtieth anniversary. Taking the same Tin Goose it had originally owned, it repainted it with the markings it had in the beginning and flew it on a nationwide tour of the United States. Pilots for the trip, which covered seventeen cities along its transcontinental route of

Northwest Airlines, were Captains L. S. "Deke" DeLong and Joe Kimm, both of whom had flown Tin Geese for the airline in the 1930s.

The tour's official opening took place on October 9, 1956, at Idlewild (now John F. Kennedy International) Airport. Parked beside the famous old Ford Tri-Motor making the tour were several other vintage planes, including a DC-3, as well as a new Douglas DC-6 and a Convair 440. After DeLong and Kimm climbed into the cockpit of the Ford Tri-Motor, the ceremonies started. "The purpose of this flight," a Northwest Orient Airlines official said, "is to illustrate the tremendous advances of commercial aviation since the airline was founded. . . . The Ford Tri-Motor has been selected because it proved that commercial air operations over the Rocky Mountains were safe."

A few minutes later the Ford Tri-Motor took off for Washington, D.C., the first stop on the cross-country venture, and for the next nine days the old plane flew westward, stopping at Detroit, Milwaukee, Chicago, Madison, Rochester (Minnesota), Minneapolis-Saint Paul, Fargo, Bismarck, Billings, Great Falls, Butte, Missoula, Spokane, Yakima, and Seattle. Large crowds welcomed the Ford Tri-Motor at every airfield, and even though the old plane showed up at its last stop at the Seattle Airport in the midst of a driving rain, hundreds of spectators were on hand to welcome the pilots and the plane.

In 1963 TWA revived its Ford Tri-Motor tour again in connection with the airline's commemoration of the Civil Aeronautics Act, which President Franklin D. Roosevelt had signed in 1938. This time the Tin Goose they flew was rented from John M. Louck, operator of the Monmouth, Illinois, airport. A retired Air Force captain who had been a World War II flight instructor, Louck had purchased the plane in 1959 after a series of swaps with several former owners in Ohio, Florida, and Indiana.

The Trimotor was a 5-AT, the seventy-fourth of that model built. First flown in September, 1929, it was originally a Ford demonstrator, and then it was rigged with floats and taken to Washington, D.C., where it was evaluated by the navy as a possible troop transport, air hospital, and torpedo bomber. Still belonging to Ford Motor Company, the plane was finally sold to Pan American Airways, which used it from 1932 to 1934. For the next three years it flew under a now unknown foreign registry, and then, in 1938, it returned to Pan Am for another three years of service. A brief stint followed with Cia Mexicana de Avancion, and then, for ten years, the plane was flown by AVIATECA in Guatemala, Central America, as a chicle cargo carrier. Finally, in 1951, it was sold to the Northwest Agricultural Aviation Corporation, Choteau, Montana. Northwest used it primarily for crop-dusting, fitting it out with "spray bars" and sending it on missions against grasshoppers in Wyoming and gypsy moths in the northwest forests.

Northwest eventually sold the plane, and it went through many ownerships. When Louck chanced upon it in Florida, it was in poor condition from its heavy wear, but after examining it carefully, he felt its wings and fuselage could be reconditioned, and he could make it serviceable. He put a lot of work into it, refurbishing the interior and installing new Pratt & Whitney engines. Once completed, the plane was sent barnstorming through the South and Midwest, its sponsor being Ford Tri-Motor Air Tours, a company that Louck organized to exhibit the plane.

Taking the plane to air-shows and aviation meets, Louck found all the passengers he wanted at three dollars a head. He took up about fifteen thousand passengers a year, some of them, ironically, pilots who had always heard of the famous old Tin Goose but who had never flown in one. At one air meet in Illinois in 1962, Louck had almost three thousand passengers. Some people waited in line four or five hours to board the plane.

When TWA leased Louck's Ford Tri-Motor for its 1963 Anniversary Tour, it flew the plane to one of its overhaul bases where it was repainted in old TWA markings; then the cross-country tour started. As on the original all-air transcontinental service, the plane flew eastward, stopping at Winslow, Albuquerque, Amarillo, and Wichita before Kansas City, where the passengers, as they had years before, stopped overnight in a hotel. Then, the following morning, the flight was resumed, with stops at Saint Louis, Indianapolis, Columbus, Pittsburgh, Harrisburg, Philadelphia, and, finally, Newark.

Only one thing remained to be done as part of the anniversary celebration—on the actual anniversary day, the plane was flown again to Washington, D.C., where "birthday" greetings from the cities visited were handed to Alan S. Boyd, Civil Aeronautics Board chairman, and Najeeb E. Halaby, Federal Aviation Agency administrator.

All in all, TWA's revival celebration had been most successful, and the publicity was excellent, the *Saturday Evening Post* describing the tour as "the last gasp of the Tin Goose."

An even bigger gasp came the following year when American Airlines introduced its 727 Astrojets. Since these jetliners were the first three-engined planes to be flown on major airlines since the Ford Tri-Motors of the early 1930s, it seemed like a "natural" publicity stunt to tour a Tin Goose to call attention—by contrast—to the new 727 Astrojet.

The Ford Tri-Motor purchased for the tour was a 5-AT owned by Hayden Aircraft, Bellflower, California, and it had once before been owned by American, although the airline had not been the original owner. After being built in April, 1929, the plane had first been purchased by Southwest Air Fast Express (SAFE), Tulsa, Oklahoma, which flew it for two years, and then it had been sold to Colonial Air Transport, Newark, New Jersey, which also flew it for two years.

In 1933 this Tin Goose was purchased by American Airways, predecessor company to American Airlines, and American used it on one of its first transcontinental flights in 1935.

Little less than a year later, when American started bringing in DC-2s and DC-3s on the run, the Ford Tri-Motor was sold to another United States airline, which shortly after sold it to TACA. For seven years TACA flew it in the Yucatan, eventually selling it to a Nicaraguan airline. From here it was bought by a Mexican freight line which operated it for several years and finally sold it to Robert Waltermire of Northwest Agriculture Station. They used it as a crop duster for a few years, and then sold it to a Mexican mine operation, which kept it for only a year before advertising in United States newspapers that the plane was for sale. An American businessman purchased it and, shortly after, sold it to Hayden Aircraft.

David A. Weiss

When American Airlines took possession of the plane on September 7, 1962, the airline issued a press release saying: "The tarnished Ford Tri-Motor landed and taxied slowly past the new ultramodern satellite buildings and continued on to our maintenance facility. Doubtless, many of the people waiting to board sleek jetliners were bewildered at the sight of this vintage aircraft. None could guess it was likely to fly again."

To promote its introduction of 727 Astrojet service in 1962, American Airlines restored a Ford Tri-Motor first flown by a predecessor company in 1933. The battered remains of the plane (below) were found in a field in Mexico; were completely overhauled (opposite, top) in American's Tulsa shops; and ended up (opposite, bottom) an authentic restored Tin Goose. AMERICAN AIRLINES

The Ford Tri-Motor was completely overhauled in American's Tulsa shops under the direction of Floyd Tohline. Twenty-four top mechanics worked on the plane, and when it was refurbished, it was an up-to-date replica of the original Ford Tri-Motor flown by American Airways in 1935. In every possible place the restoration had been authentic, but, in a few areas, modern touches—like new improved engines and variable-pitch propellers—had been added.

American took the plane on a nationwide barnstorming tour where it was seen by three million people and flown in by ten thousand. Accompanied by a 727 Astrojet, the old Tin Goose proved to be just as big an attraction. Whether it was courtesy rides, or just being inspected by the curious, the Ford Tri-Motor proved a potent draw.

American also flew the plane in connection with specific stunts. At the Democratic Convention of 1964 it flew to the convention a special party headed by Representative James R. Roosevelt, in commemoration of the flight that Roosevelt's father—FDR—had made from Albany to Chicago in 1932 to accept the presidential nomination.

Shortly after, American Airlines announced it would renovate the plane again and offer it to the Smithsonian Institution's National Air Museum. The renovation would consist of bringing the plane back closer to its original condition. The new Pratt & Whitney engines would be replaced by less powerful Pratt & Whitneys similar to the Wasps first used on the plane. The variable-pitch propeller would be discarded in favor of an old-time propeller. When the National Air Museum would be ready for the plane, American would ship it there to take its place along with other great planes like Lindbergh's *Spirit of St. Louis* and the Wright brothers' original airplane.

Anticipating the eventual loss of the Ford Tri-Motor donated to the National Air Museum (now the Smithsonian Air and Space Museun)), and realizing the publicity value in one of the ancient Tin Geese, American in 1965 purchased another Ford Tri-Motor for exhibition at the New York World's Fair. The seller was Louck; the plane, the same one he had rented out previously to TWA for its 1963 anniversary tour.

The plane made headlines when one night a group of juvenile delinquents broke into American's exhibit at the Fair and supposedly "wrecked" the plane. It was discovered later that all the dents and bruises found on the Trimotor's fuselage had been inflicted, not by destructive youth, but, rather, by the rigors of flying for years in the Guatemalan jungle.

Chapter Twelve
Still Flyin'

It would seem as the aviation world neared the remaining years of the 20th century the Tin Goose story was fast coming to a close. Although rumors sprang up from time to time that an old Tin Goose was here or there, disassembled and packed away in some old airport's hangar or farmer's barn, few took them seriously. Neither did they believe any Tin Gooses could still be flying in South or Central America. News of the old planes occasionally popped up in aviation magazines, only to vanish soon after and then reappear several years later.

But unbelievably in the 1960s a small group of aircraft engineers started working to revive the Tin Goose and manufacture a modernized version of the plane for use in bush and STOL (Short Takeoff and Landing) operations.

Ever since 1933, when Henry and Edsel Ford stopped manufacturing Ford Tri-Motors, the aviation industry had talked about someone, hopefully the Fords, would resume production of the planes. After all, if the Tin Geese were so good at bush operations, certainly there was need for more planes than those surviving from the original 199. After all, too, were not 90 percent of the world's airports really bush airports, infinitely more suited to these slow, lumbering planes than to their sleek, speedy successors?

A few years after World War II, Northrop Aircraft announced a new plane, called the Pioneer, to be constructed along the design of the old Ford Tri-Motor, and it ended up manufacturing thirty of them. The announcement was particularly ironic since Northrop was the builder of the famous Flying Wing, startlingly similar in design to Stout's original Batwing plane of three decades earlier. But nothing ever came of Northrop's plans.

In the mid-1950s, however, another such project was announced. A group of six aeronautical engineers working for Douglas Aircraft and other airplane manufacturers in the southern California aviation and manufacturing complex began investigating the possibilities of re-producing the old Ford Tri-Motor. Headed by Robert E. Hayden, this group formed Hayden Aircraft in 1954 to see what could be done about it. For several months no more was heard about the company, but the following year Hayden Aircraft issued an announcement that proved to be a news bombshell: the director of engineering for the new Hayden Aircraft Corporation would be none other than William B. Stout, the man who, together with Henry and Edsel Ford, had developed the original Ford Tri-Motor.

Opposite: The Bushmaster 2000—the 1960s version of the Ford Tri-Motor. AVIATION WEEK AND SPACE TECHNOLOGY

By this time Stout was the only one of the original developers still alive. Edsel had died in 1943, just as the Ford Motor Company was in the midst of its gigantic program of producing B-24 Liberator bombers at Willow Run for the United States war effort. His death at the age of 49 was tragic. A victim of stomach cancer, for which he was being treated, he is thought to have succumbed as a result of undulant fever contracted from milk obtained from the Fords' own farm. A firm adherent of old-time farming methods, Henry Ford did not believe in pasteurization.

Edsel's role in the growth of the Ford Motor Company did not go unrecognized by the press. "He was a full partner of his father, not merely a collaborator, but an originator," *The New York Times* wrote in an editorial. Also recognized was Edsel's role in the company's aviation activities. "We must give him credit for the Ford Motor Company's war effort at River Rouge and the bomber plant at Willow Run where 8,000 Liberator bombers have been turned out," the *Times* editorialized. "It was he who brought about the association of the company and the enthusiastic enlistment of Henry Ford in the business of aviation."

Henry Ford was stunned by the premature loss of his only child, but, never one to shirk a responsibility, he resumed at the advanced age of seventy-nine the active management of the company that he had relinquished to Edsel a quarter century before. He also wisely began grooming his grandson and namesake to take over control of the company. When, four years later, Henry Ford died, Henry Ford II was running Ford Motor Company.

Stout had kept himself busy in the years after he had left Ford Motor Company. Reestablishing his own Stout Engineering Laboratories, he continued experimenting in not only the aeronautical field, but also in the automotive, railroad, and half a dozen other fields.

Stout's laboratory headquarters was a large factory building on Telegraph Road in Dearborn, not far from the Ford Motor Company, and here, in an office from whose ceiling hung a model of the famous Ford Tri-Motor, Stout ironically turned his attention to two projects similar to those that Henry Ford had pushed in the early days of the Airplane Manufacturing Division and that Stout had not been particularly enthusiastic about. One of these was a fifty-passenger metal passenger plane. The other was a duralumin Skycar flivver featuring a rear-engine propeller and a four-wheel landing gear that enabled the craft to be let down, as Stout explained it, in a basket-of-eggs fashion.

When Stout tried to get financing to get these projects off the drawing board, he found no money available. The nation was still in the throes of the Depression, and no one felt like making an investment in such revolutionary aircraft. Undaunted, Stout now turned to automobiles where, as he put it, he "continued his dogfight with the past." Looking at the auto industry objectively —unhampered by conventions and production lines—Stout could see only an industry restricted by its huge investment in machinery. As he viewed the auto industry, too much of its ingenuity was spent developing tools that would cut a cent or two out of the total production cost. "It's like inventing rubber gloves for leaky fountain pens," he told a magazine reporter, "and the end-product is a 'modern' automobile as obsolete as a horse-drawn vehicle."

To show what could be done, and to challenge the automobile industry to renounce its conservatism and to abandon traditional design, Stout in 1936 came out with the first of his famous Scarab experimental cars.

Shaped like a beetle (it looked almost the same from the rear as it did from the front), the Scarab was as revolutionary in the automotive field as Stout's first all-metal Batwing planes had been for aircraft.

The American automobile industry later adapted many Scarab features, such as eliminating running boards to make for greater interior width. As Stout explained his concept, "Running boards are only there because buggies had them." But other Stout ideas remain to be accepted. The Scarab's back seat, for example, could be used as a couch, and next to it was a collapsible table. Also, the auto's engine was installed in the rear ("it's logical, since that's where the weight is"). Stout claimed that, as a result, the car skidded less, it steered better, its floor could be lowered and its passenger area increased (because no drive shaft was needed), and that, around turns, the Scarab would bank instead of roll. To prove the superiority of the Scarab, Stout—never at a loss to dream up a good publicity stunt—drove the car from Detroit to California with a glass of water on the dashboard. In the entire trip of several thousand miles, the water never spilled.

The Scarab had other unusual features too: rubber-filled upholstery, an air-conditioning and ventilating system, a slanted windshield for greater visibility, and no hood. All these features, as well as Stout's down-to-earth approach in designing the famous car, received extensive publicity, "Never resort to mathematics until you have exhausted the possibility of two toothpicks and a piece of string," Stout once told a Detroit engineering society, and he proved the practicality of this approach many times in designing the Scarab. Later on, the auto industry adopted Stout's innovation in the Scarab of having push-button doors. The Detroit auto companies spent thousands of dollars developing the button mechanisms. Stout, for the Scarab, installed seventy-five-cent mechanisms he bought at a hardware store, and they worked just as well.

The first 1936 version of Stout's revolutionary rear-engined Scarab automobile,
UNITED PRESS INTERNATIONAL PHOTO

The Scarab car was offered as a custom car, and it was advertised in *Fortune* magazine as "a challenge and prophecy," the advertisement saying that a demonstration of it was "upon invitation only." Stout was able to produce the cars for sale at $5,000, but only nine were ever sold.

Stout turned next to railroad design. The Pullman Car and Manufacturing Corporation gave him a blank commission to design from scratch a new type of passenger train, and they promised him no interference from the railroad engineers on their staff. The result was the revolutionary Railplane, into which Stout constructed many of the design features he had pioneered for the Ford Tri-Motor. Constructed of welded tubing covered with duralumin, the Railplane was the first lightweight, high-speed, streamlined train in the United States. Its cars were soundproofed and air-conditioned. They were also silent, because they rode on rubber wheels. Most ingenious of all, the train was driven by standard gasoline engines that could be repaired by any regular truck service.

The Pullman Corporation was impressed with the train's speed of 90 mph and the comfort offered its passengers, but the train's lightweight duralumin construction produced an unenthusiastic reaction from railroad men reared on the theory that strength comes from weight. No sooner did Pullman receive the Railplane from Stout than it shipped the train to a Chicago warehouse, where it has reposed ever since.

Having unsuccessfully tried to innovate in the automotive and railroad fields, Stout now returned to aviation. In 1941 he introduced a new version of his Skycar air flivver. The original one produced in the 1930s had been constructed of duralumin. This one was built of stainless steel. Stout had decided upon stainless steel, he said, because it was more corrosion-proof than duralumin. Also, unlike duralumin, which had to be riveted, stainless steel could be spot-welded at a fraction of the cost.

A two-seater, high-wing monoplane with its engine and propeller in the rear, the Skycar was a "fool-resistant" plane. With special two-speed gear, the plane could take off in a space less than two hundred feet long. For landing, it had four wheels, so that nosing over was almost impossible. Stout envisioned the plane selling for no more than $1,000, and said it would go into production as soon as he could obtain a 100-hp engine that would weigh no more than one hundred pounds and cost no more than one hundred dollars. But, alas, no engine was ever obtained, and Stout's Skycar—"the world's first motor car of the air"—never went into production.

Still undaunted, Stout two years later announced his "Heli-car," a combination automobile-helicopter. A car with detachable wings, this strange vehicle was designed to enable a vacationing family to hop over traffic, bodies of water, and roadless forests. But it too was never produced.

After World War II, creative inventions continued to flow from Stout's fertile brain. In 1946 he designed a third version of his Skycar, this one constructed of fiberglas. About the same time he announced an all-metal super-trailer that could be collapsed and telescoped for transport to another locality, and then, upon arrival, extended and converted into a beautiful home. "I think it's one of the answers to the housing problem," Stout said.

An air-conditioned bed, a theatre seat that automatically pushes back to let people pass—Stout continued his experiments, first in his Stout Engineering Laboratories in Dearborn, then later in Phoenix, Arizona, where he moved the laboratories in the early 1950s.

Stout workingon a model of his Skycar air flivver in 1941. UNITED PRESS INTERNATIONAL PHOTO

When the call came from Hayden Aircraft, Stout was only too happy to become involved in the revival of the famous Ford Tri-Motor. After the announcement was made to the aviation press that he would become Hayden's director of engineering, the aviation press directed to Stout the skeptical question that everyone was asking: "Why, in an age of supersonic jets, would anyone consider reviving a lumbering old transport whose top speed was 150 mph?" one editor asked.

Stout, as always, had an answer. "The Ford Tri-Motor was designed to be the super passenger airplane of its day," he said. "But unrealized by us who designed it, the plane was also unbeatable in certain types of freight operations, and that is why we feel this plane has a definite market today," he said.

Robert E. Hayden, president of Hayden Aircraft, also had an answer for the aviation press. "We have organized our company in the recognition that no aircraft in production today adequately fills the need for a simple, low-cost, large-capacity, multi-engined craft capable of operating from short, rough, and high-altitude landing strips in vast underdeveloped regions of the world."

Hayden made further announcements as the company's plans developed. The plane to be built along Ford Tri-Motor principles would be called the Stout Bushmaster. It would be a grass-runway, freight-passenger utility airplane with high lift and short takeoff and landing capabilities. Hopefully the Stout Bushmaster would be able to carry a gross load equal to its own empty weight (6,500 lbs.). When fully loaded, it would be able to take off with 600 feet of ground roll and clear 50-foot obstacles within 1,000 feet. Maximum speed would be 150 mph, cruising speed, 125 mph, and service ceiling, 20,-000 feet. The cost of the plane would be between $100,000 and $125,000, "What we will have in the Stout Bushmaster," Hayden said, "is an airplane that has the load-carrying capacity of the DC-3 combined with the capability to get in and out of the same places as a Piper Cub."

Hayden Aircraft had high hopes for the Stout Bushmaster. They saw it useful for not only passenger and freight service in remote areas, but also in conducting aerial surveys and explorations, in flying short feeder routes (such as the Island Airlines route), in dusting crops and aiding in forest conservation, and as a military transport for the use in areas of rugged terrain. "No plane now available can perform as we expect the Stout Bushmaster to perform, and still carry the load it does," Hayden said.

In the initial planning for the Stout Bushmaster, it was decided to add many new features to the old trimotor design. The engine power would be doubled by using new Pratt & Whitney 450 hp R-985 engines. The fuselage would be widened, resulting in larger cargo space. A large jeep-sized cargo door would be added. Passenger capacity would be increased from fourteen or seventeen to twenty-one. The floor of the plane would be strengthened and the cockpit would be modernized, with many new types of instruments being added. Hydraulic struts would also be installed in the landing gear, and external cables would be rerouted internally.

Still, the overall design would be based on the Ford Tri-Motor, and to facilitate the drawing-board stages of the project Stout wrote Henry Ford II for permission to examine the original drawings and engineering data still in the possession of the Ford Motor Company. Henry Ford II not only gave permission, but generously sold all the plans to Hayden, giving Stout exclusive use of them on the condition that the name of Ford not be used in advertising and publicizing the new plane.

Stout got to work, and within a few months Hayden Aircraft had about sixty people working on the design of the Stout Bushmaster. Most were engineers employed in full-time jobs with other aircraft manufacturers in the area. They would spend their evenings and weekends at Hayden, their work being compensated for by Hayden stock. After the Stout Bushmaster was in production, they expected to work full time at Hayden.

Stout had estimated that Ford's furnishing of the original plans saved about two and a half million dollars' worth of engineering man-hours, but, unfortunately, when he got deeper into the drawings, he found that some key ones were missing. At this point, Stout came up with another of his seemingly crackpot ideas. Why not locate an old Ford Tri-Motor and purchase it just for the purpose of copying first-hand its components and dimensions?

It was not easy to find an available Tin Goose, but after months of searching, one was found in Mexico. The old plane was sitting on concrete blocks, and a Mexican family of five was living in it, but the plane was intact. Its wing, engines, controls—everything was in order.

Hayden had to go through some fancy business dealings to acquire it, but the company was finally successful. Put into operable condition, the old Tin Goose was flown to Long Beach Airport. Here, piece by piece, it was taken apart and put back together again. In this process, Stout obtained all the detailed information he needed to complete the drawings for the Stout Bushmaster.

While this work was going on, the Hayden Ford Tri-Motor was frequently used for barnstorming. Once an air show held in Prescott, Arizona, invited Stout to fly in the plane. Accepting, the old inventor flew down with his Hayden Aircraft associates. Arriving in Prescott, they surprised the crowd by ignoring the runway and landing the Tin Goose in a nearby cow pasture. Despite the many planes at the show, most of the crowd rushed over to see the ancient Ford Tri-Motor. It was the hit of the air show.

It had been planned for Hayden Aircraft to complete its prototype of the Stout Bushmaster in 1957, but suddenly, on March 20, 1956, Stout died at the age of seventy-six.

Hayden Aircraft tried bringing in new executives. Rex Noville, Admiral Byrd's onetime executive officer who had flown Ford Tri-Motors in the upper Amazon, joined the company and became chairman of Hayden's advisory board. But after a while the project languished. Even the old Ford Tri-Motor that the company owned was sold to American Airlines.

By the early 1960s, however, interest revived. Robert Hayden himself was no longer involved. The two prime movers now were Laurent E. "Frenchy" Savard, an old Ford Tri-Motor pilot and one of the original Hayden Aircraft stockholders, and Ralph Williams, president of Aircraft Hydro-Forming Corporation of Gardena, California, a company that manufactured components, like leading edges and bombing racks.

Williams bought out Hayden's interest in the Stout Bushmaster and started up actively again. "There are still lots of skeptics," he told the press. "But we are determined to get this project on the road. We still believe that the basic Ford design has a role in the new air age, and are more convinced than ever that such a plane can do some jobs better than any other aircraft."

Aircraft Hydro-Forming Corporation began the new project by conducting a survey to see if it really had a market for a new Ford Tri-Motor. Inquiries were made of several hundred small airlines and, much as Williams had predicted, interest was shown by companies all over the world, from New Guinea and New Zealand to Madagascar and Alaska.

The next step was to complete the drawings for the new plane. Actually, in the years that had elapsed since Hayden had suspended the project, it had been decided that modifications would have to be made in the Stout Bushmaster as Stout had conceived it. So a slightly different plane would now be constructed along old Ford Tri-Motor lines, and it would have a new name—the Bushmaster 2000.

Like the Stout Bushmaster, the Bushmaster 2000 would be a modern version of the old Ford Tri-Motor. It would have the same all-metal construction of the trimotor, but the metal skin used now would be lighter and stronger. Other changes would also be made, many originally proposed in the Stout Bushmaster. It would have a king-sized door big enough to take on heavy equipment and to offer egress to parachutists. It would have Pratt & Whitney R-985 engines. It would have three-bladed, full-feathering propellers.

The old Ford Tri-Motor barn door-like rudder would be replaced by a dorsal-fin rudder that would give the plane better directional stability at low speeds. The cabin would be roomier; the cockpit more modern, with yoke-type controls; the instrument panel would contain the latest flight and navigational instruments; and the control cables would be inside.

Drawings for the new Bushmaster 2000 were completed at Aircraft Hydro-Forming Corporation by squeezing the work in between a thousand other projects, and laboring long and hard at nights and on weekends. Once the drawings were completed, the more formidable task remained of building a prototype of the plane. This was accomplished under even more difficult conditions. "We bootlegged time, labor, and machines," Savard said later.

David A. Weiss

The plan had been to assemble the smaller components at Aircraft Hydro-Forming Corporation's Gardena plant, and then to make the final assembly at Long Beach Airport. In August, 1966, the plane's components were ready to be towed to Long Beach, At 3 A.M., to avoid traffic, the completed fuselage and inboard wing panels were trucked down the freeway, the engines and wing following. In a big hangar at Long Beach Aero Center the Bushmaster 2000 was put together.

Out of curiosity, hundreds of pilots and aviation enthusiasts came out to look at the revised edition of the famous old Ford Tri-Motor. They easily noticed the changes: the different rudder shape, the big door, the new engines, the additional streamlining. But when they saw the plane waddling out to the runway, all they could say was, "My God, it's a new Tin Goose."

The test pilot engaged to fly the new plane was William B. Bridgeman, the former Douglas test pilot famous for flying the D-558 Skyrocket at Edwards Air Force Base. Bridgeman, who would be tragically killed several years later in a seaplane crash off Catalina, took up the Bushmaster 2000 and flew it to Chino Airport, fifty miles to the north. His passengers on the test flight were Frenchy Savard and Bob Lanley, president, Catalina Channel Airlines.

The Federal Aviation Agency certified the Bushmaster 2000, and Aircraft Hydro-Forming Corporation soon published a flight manual for the plane. In this manual, its top speed was listed as 145 mph; minimum control speed, 63 mph; best climb speed, 73 mph; and normal operating speed, 129 mph. Takeoff distance (to clear a 50-foot obstacle) was 1,200 feet, and landing distance (again to clear a 50-foot obstacle) was 1,725 feet. "What we have here," Williams said, "is a plane designed to be a workhorse. On any two of its three engines, it can take off, cruise, or land. It can carry a big pay-load, can take off or land fully loaded on a runway only the length of two football fields, and it can easily be adapted for floats, skis, or other combinations of equipment."

Aircraft Hydro-Forming Corporation planned to build one more prototype, and then it expected to go into production, manufacturing, he

264 |

said, three Bushmaster 2000s a month. Meanwhile, the completed plane would be sent on an international tour to demonstrate its capability to prospective customers. The publicity obtained from the first Bushmaster 2000 was considerable, and, as a result, hundreds of inquiries came from all over the world from individuals and companies seeking particulars about the plane. The United States Forestry Service expressed interest. The United States Army explored its utility, seeing it as a possible transport for use in Vietnam. Several delegations from South America, which had traveled to California to see the Bushmaster 2000, were also interested.

According to the company, the original Bushmaster 2000 after seventy hours of flying was sold to "an American corporation," but this does not seem to have been the case. It was eventually given to a local air museum in Oakland. Meanwhile, the corporation was building the second plane, redesigning it in the hope of reducing its weight five hundred to a thousand pounds. "We are beginning to think the heavy reliance on stress work done for the Ford Tri-Motor was excessively conservative," he said. "Thus if possible we want to lower the weight of the Bushmaster 2000 from 8,544 pounds to 7,500 pounds."

The immediate target of the weight reducers was the plane's heavy multi-spar wing. "By redesigning this, we can introduce lighter-weight construction," Williams said. Other improvements planned besides weight reduction included upgrading the plane's braking system with double-disc brakes in order to reduce landing-roil-out distances.

By November 1969, the new designs were being applied to the second Bushmaster 2000 under construction, and soon after, the company underwent another reorganization. Spun off from Aircraft Hydro-Forming Corporation was Bushmaster Aircraft Corporation, with headquarters at Long Beach Municipal Airport. Here the company began manufacturing more Bushmaster 2000s, and by 1971 supposedly had three other Bush-master 2000s under construction. Williams, still president, was as confident as ever about the plane's future. "We've reduced the weight," he said, "and now we think we have an excellent weight-to-payload combination that will make the Bushmaster 2000 most attractive to potential purchasers." According to Williams, the break-even point for the company was only twenty Bushmaster 2000s. With prices ranging from $175,000 to $200,000, depending on the operational items desired, he said he ranticipated far greater sales.

Chapter Thirteen
The Tin Goose In
The Space Age

Not for more than a quarter century have Ford Tri-Motors made their scheduled flights for Island Airlines between the mainland and Put-in-Bay Island in Lake Erie. Not only did Island Airlines sell off all its Tin Gooses in the 1970s and 1980s, but the airline itself is no more, having gone out of business in 1992.

No longer too are barnstormers flying from city to city showing off their Ford Tri-Motors and giving 15-minute rides to townspeople. John Louck who revived barnstorming in the 1960s gave up by the end of the decade, selling his Tin Goose to American Airlines. Gaylord Moxon, who ran a classy barnstorming operation on the West Coast—tickets were sold by aides in red vests—also quit before 1970. So did Charles LeMaster. He and his wife had barnstormed around the Midwest for five years in his *Kansas Clipper,* but he sold it in 1977 to Grand Canyon Airlines, explaining that the barnstorming circuit had taken them away from home at least half of every year.

Last of the barnstormers was Al Chaney who barnstormed in a Tin Goose painted red, white and blue, He gave up in 1991.

As for the Bushmaster, that "modern" version of the Ford Tri-Motor, it never got off the ground despite all the rosy predictions. It was not until 1966, eleven years after the company started to build its first plane—they named it the Stout Bushmaster 2000—that its construction was completed, and eleven years after that for the second one called Bushmaster 2000 to see the light of day. A third Bushmaster had beeen started, but it was never finished.

Fly in a genuine 1929 Ford Tri-Motor!

Sept 6 – 9, 2012	Mankato, MN Mankato Regional Airport	Sep 6 – 2 – 5 P Sep 7 – 9 – 9 A
Sep 6 – 8, 2012	Galesburg, IL Galesburg Regional Airport	Sep 6- 8, 2012 : 5:00 PM
Sep 13 – 16, 2012	Sioux City, IA Sioux Gateway Airport	Sep 13– 2 – 5 F Sep 14 – 16 – 9
Sep 13 - 16, 2012	Greencastle, IN Putnam County Airport	Sep 13 – 2-5 Pt Sep 14-16 9AM
Sep 20-23, 2012	Bloomington, IL Central Illinois Regional Airport	Sep 20 – 2-5 Pt Sep 21 – 23 9 A
Sep 20-23, 2012	Racine, WI John H. Batten Airport	Sep 20 – 1-5 Pt Sep 21 – 23 9 A

Dates are subject to change due to weather, mechanical, etc.

Chapter Application for EAA Ford Tri-Motor Tour
Application deadline for **2013**

9/10/2012

One of the EAA flyers announcing the 2012 Tour

The Stout Bushmaster 2000 after many years in Thomas J. Watson, Jr.'s Owls Head Transportation Museum is now in the Golden Wings Flying Museum in Blaine, Minnesota and is sometimes flown around that area by its owner Greg Herrick. As for the Bushmaster 2000, it flew for a while over Denali National Park in Alaska, but its life ended when it crashed on take-off in 2004 in Fullerton, California due to an error on the part of its mechanic.

A final blow to the overall project was the consensus that although the Bushmaster was not a total failure, it was not any real improvement over any number of other planes designed for similar service, and most critics agreed that its production was also the victim of poor management, poor financng and poor marketing.

As for the number of Ford Tri-Motors that have survived, the number—it is eighteen—is not much diferrent than it was a decade or two ago. A rcent suvey reveals that of the 199 Ford Tri-Motors manufactured eight are flyable, five non-flyable, and another five in various stages of restoration.

The most active Ford Tri-Motor today is the 4-AT owned by the Experimetal Aviation Association (EAA), an international organization of some 160,000 aviation enthusiasts and more than 800 chapters which is famous for its summer "Fly-In" when 15,000 (yes, 15,000.) airplanes and pilots convene at its Oshkosh, Wisconsin headquarters. EAA has an AirVenture Museum there which houses more than 150 antique airplanes and in an extension, a tram ride away, is Pioneer Airport, a replica of a 1930's airport, which has fifty more vintage airplanes stowed in seven hangars, the first of which contains a 4-AT Tin Goose which is Pioneer's chief attraction.

How this Tin Goose came to Pioneer Airport is a story in itself. Originally purchased in 1929 by what became Eastern Airlines, it flew Eastern's major routes for a few years before being leased to Cubana Airlines where it started up passenger service between Santiago and Havana. Then after flying in several other Caribbean countries it was returned to the United States where under several different owners it was refitted for use as a crop duster. With two new 450 HP engines and one 550 HP engine added it became the most powerful model 4-AT ever flown. In 1955 the plane t was moved to Idaho and fitted with two 275 gallon tanks and bomb doors for use as a borate bomber in aerial fire fighting. Then in 1958, it was further modified for use by smoke jumpers.

After working for a variety of crop spraying businesses, the Tri-Motor then moved to Lawrence, Kansas where a new owner used it for barnstorming tours. During this period it also served as the primary setting for the Jerry Lewis comedy, "The Family Jewels." In 1973, the aircraft was still being used for air show rides, but at the annual 1973 EAA Fly-In, a severe thunderstorm ripped the plane from its tie-downs, lifted it 50 feet into the air and smashed it to the ground on its back.

It took thirteen years with the assistance of Hov-Aire's Maurice Hovious, the world's greatest expert on restoring Ford Tri-Motors, to restore the Tin Goose, and when this was completed, the plane was moved to Pioneer Airport where during the summer months it offered short rides at a cost of $80 per trip for 15-minute flights around Oshkosh.

One thing to led to another and soon EAA was offering one- and two-day trips in the Ford Tri-Motor, a program that was expanded further in 2003 when EAA launched the first of its Grand Tours, lasting almost a month and taking in some ten cities, most of which were in the mid-west. Then, most recently—in 2012—EAA added a second Grand Tour at the same time with a second Ford Tri-Motor borrowed from the Kalamazoo Aviation History Museum's AIRZOO. ZOO.

AIRZOO which lent its Ford Tri-Motor to EAA for the EAA 2012 Tour was founded in 1997as Kalamazoo Aviation History Museum. Located next to the Kalamazoo, Michigan-Battle Creek International Airport, it now includes an indoor amusement park as well as an aviation museum. Through the years the museum expanded quickly, and in 1990 it added a special building which contains on the subject of flight the longest mural in the world as well as special research and educational facilities and a plane restoration center. Today it is one of the ten largest non-governmental aviation museums in the United States, holding many historical and rare aircraft, including the world's fastest air-breathing aircraft, the SR-71B Blackbird.

AIRZOO's 4-AT Tin Goose like many of its vintage planes is flyable. For years it had flown for Johnson Flying Service until it cracked up in 1959, After Johnson merged with Evergreen International, the giant air freight corporation, Evergreen sold the plane to the Museum who turned it over to Kal-Aero, a company that specialized in plane restoration. Progress was slow. Originally the restoration was to go only as far as to make the plane suitabl;e for exhibition, but after a while it was decided to go all the way and make it flyable. A new timetable was set which called for completion of the restoration in time for the 1991 EAA annual convention. Three crews worked night and day to get it ready. They succeeded, and Evergreen's Penn Stohr was brought in to fly the plane from Kalamazoo to Oshkosh where it won a special EAA award.

Afterwards Evergreen donated the plane to AIRZOO where it now participates in air shows, and is loaned on special occasions.

Evergreen itself has long had an aviation museum. Located at the company's headquadquarters in Minnville, Oregon, the AirVenture museum contains many antique and historic planes, including not only a Ford Tri-Motor, but, among others, a Fairchild, WACO, DeHavilland, and Stinson in addition to its major airplane exhibit the famous Spruce Goose, constructed of wood by Howard Hughes during World War II and still the largest plane ever built.

The Tin Goose owned by Evergreen was originally a TWA plane that had found its way to Mexico where it had been overhauled with its corrugated skin replaced by sheet duraluminum, giving it the distinction of being the one and only "smooth-skin" Ford Tri-Motor. The plane cracked up in 1954, after which it was purchased by Harrah's Auto Museum which restored it beautifully to its original skin and configuration. In 1990, when the Harrah's collection was broken up, the plane was purchased for $1,500,000 at auction by Evergreen.

Pilot of the Evergreen Ford Tri-motor was Penn R. Stohr, Jr., whose father, Penn Stohr, was a famous bush pilot who was killed flying a Tin Goose for Johnson Flying Service, the company that did crop dusting for many years for the U.S. Forestry Service. When Evergreen acquired Johnson Flying Service in 1979, Penn Stohr, who had been an executive with the company, became Evergreen's senior vice president in charge of flight operations. He flies to air shows from time to time and to special events like reunions of smoke jumpers. As part of EAA's summer "Fly In" there are activities for "young eagles" who get to "put hands on" the planes as adult pilots fly them. One year Penn Stohr's twin children participated in this program, making three generations of Stohrs associated with Tin Gooses. Meanwhile, the plane is on exhibit at Evergreen's AirVenture museum, and although it is flyable, there are no plans to move it out of static display.

Two of the flyable Tin Gooses are owned by Greg Herrick, president of the Aviation Foundation of America and the Antique Airplane Association. Both are AT-4s, one in Jackson, Wyoming, where Herrick lives, and the other in Herrick's Golden Wings Flying Museum at the Anoka County-Blaine Airport near Minneapolis. Minnesota,

This Tin Goose has many claims to fame. For one thing, manufactured in 1927—the tenth Ford Tri-Motor to be produced—it is the oldest Tin Goose in existence. For another, it inaugurated commercial flight between United States and Mexico, and it was also the first passenger airliner plane to fly over the Rockies.

Another unusual fact about this Tin Goose is that it was the poster boy for the entire line of Ford Tri-Motors; its photograph was used in all the advertising printed in support of the Ford Tri-Motor sales effort.

The plane also made headlines in 1927 when at William B. Stout's suggestion Henry Ford aarranged to fly Mrs. Evangeline Lindbergh to Mexico City on Christmas Eve to surprise her famous son who had flown there for the holidays.

The original owner of this Tin Goose was Sky View Airlines which used it for several decades to fly tourists over and aroundNiagara Falls—the reason the plane was given the name "|Niagara". Sky View Airlines went bankruprt during the Depression, after which the plane was purchased by a pilot named Grant McConachie who used it for bush operations in North Canada. An accident at Telegraph Creek in 1941 cracked its fuselage and shortly after, the plane was grounded by Canadian air authorities and abandoned at Carcross in the Yukon. Fifteen years later Gene Frank in Caldwell, Idaho, purchased the wreckage, and trucked the big pieces down the Alaska Highway to Caldwell, only to return to the Yukon in 1983 to pick up the remaining pieces.

Greg Herrick in the meantime had a highly successfulful career as the owner of ZEOS International, a computer manufacturing company which he sold out for $60 million and is now publisher of the mail-order company Historic Aviation as well as the founder and owner of the Golden Wings Flying Museum with its collection of vintage airplanes where he concentrates on what he calls the Golden Age of Airplanes, planes from the 1920s and 1930s.

All in all he has seven trimotors in his collection and how he obtained the Tin Goose is a story in itself. It started with Gene Frank whom he had approached for years offering to purchase the remains of Frank's Tin Goose, but Frank held out because he intended to restore the plane himself, but Herrick was persuasive, pointing out that Frank was eighty years old and that he (Herrick)was in a better position to complete the restoration faster. Frank finally gave in, and Herrick immediately turned over the restoration to Maurice Hovious.

Herrick besides collecting vintage aircraft frequently recreated famous flights with one or more of his antique planes. In 2001 he restaged one of Amelia Earhart's intercontinental flights and two years later organized National Air Tour just like the old Ford Reliability Tours of the 1920s-1930s. Sponsored by the Ford Motor Co., the original sponsor of three decades before, it consisted of thirty-five vintage arcraft, including the restored Ford Tri-Motor, traveling 4,000 miles and visiting twenty-six cities in the mid-west.

Like most of the planes in the Golden Wings Flying Museum, the Ford Tri-Motor today is not only flyable, but available for open houses, personal appearances, breakfast events, air shows, and as props in movies.

Another flyable Ford Tri-Motor is in Kermit Weeks' Fantasy of Flight. Located In Polk City, Florida, not far from Disney World, it is the world's largest private collection of historic aircraft containing over 150 planes. Like many aviation enthusiasts Weeks got interested in aviation at an early age. When only seventeen with a friend he started to build from scratch a home-made airplane which actually flew when they completed it four years later. His first expertise in aviation was in aerobatics where he successfully competed internationally in addition to winning two United States championships. Iin 1986 his interest turned to acquiring and restoring historic planes, and with his own collection of planes, opened Weeks Air Museum in Miami, Florida. When Hurrcane Andrew struck in 1992,the museum and the planes suffered severe damage, and although Weeks was able to re-open the Miami museum two years later, his emphasis shifted to Polk City where he had built a vastly bigger museum which opened in 1995 as Fantasy of Light.

The Ford Tri-Motor in Fantasy of Flight—it is the "The City of Philadelphia", a 4-AT.—was the plane featured in the film "Indiana Jones and the Temple of Doom" released in 1968. More important, it was this plane that in 1929 inaugurated transcontinental air passenger service when it pulled in the Glendale, California, airport with Charles Lindbergh at the controls and screen star Gloria Swanson on hand to do the christening honors with grape juice.

The story of Weeks's Ford Tri-Motor illustrates how some Tin Gooses are revived, almost phoenix-like, from what would seem like the end of the line. It had been owned for years by Island Airlines and then in 1977 after it was damaged in a storm, it was sent to Kal-Aero, an airplane maintenance and repair company in Kalamazoo, Michigan which was able to restore it after three years and returned to Island Airlines which after a few years sold the plane to Kermit Weeks for his Weeks Air Museum only to have seriously damaged again by Hurricane Andrew in 1912.

This time it was sent to Battle Creek, Michigan, to Hov-Aire, another company specializing in the restoration of damaged planes. By coincidence, proving that fact is stranger than fiction, the president of Hov-Aire was Maurice Hovious, the same man who, while working for Kal-Aero, had repaired the plane in 1977 when Island Airlines owned it.

Two of the flyable Tin Gooses are owned by Greg Herrick, president of the Aviation Foundation of America and the Antique Airplane Association. Both are AT-4s, one in Jackson, Wyoming, where Herrick lives, and the other in Herrick's Golden Wings Flying Museum at the Anoka County-Blaine Airport near Minneapolis. Minnesota,

This Tin Goose has many claims to fame. For one thing, manufactured in 1927—the tenth Ford Tri-Motor to be produced—it is the oldest Tin Goose in existence. For another, it inaugurated commercial flight between United States and Mexico, and it was also the first passenger airliner plane to fly over the Rockies.

Another unusual fact about this Tin Goose is that it was the poster boy for the entire line of Tri-Motors; its photograph was used in all the advertising printed in support of the Ford Tri-Motor sales effort.

The plane also made headlines in 1927 when at Stout's suggestion Henry Ford aarranged to fly Mrs. Evangeline Lindbergh to Mexico City on Christmas Eve to surprise her famous son who had flown there for the holidays.

The original owner of this Tin Goose was Sky View Airlines which used it for several decades to fly tourists over Niagara Falls—the reason the plane was given the name "Niagara". Then, when Sky View Airlines went bankruprt during the Depression, the plane was purchased by a pilot named Grant McConachie who used it for bush operations in North Canada. An accident at Telegraph Creek in 1941 cracked its fuselage and shortly after, the plane was grounded by Canadian air authorities and abandoned at Carcross in the Yukon. Fifteen years later Gene Frank in Caldwell, Idaho, purchased the wreckage, and trucked the big pieces down the Alaska Highway to Caldwell, only to return to the Yukon in 1983 to pick up the remaining pieces.

Greg Herrick in the meantime had a highly successfulful career as the owner of ZEOS International, a computer manufacturing company which he sold out for $60 million and is now publisher of the mail-order company Historic Aviation as well as the founder and owner of the Golden Wings Flying Museum with its collection of vintage airplanes where he concentrates on what he calls the "Golden Age of Airplanes," planes from the 1920s and 1930s.

He has seven trimotors in his collection and how he obtained the Tin Goose is a story in itself. It started with Gene Frank whom he had approached for years offering to purchase the remains of Frank's Tin Goose, but Frank held out because he intended to restore the plane himself, but Herrick was persuasive, pointing out that Frank was eighty years old and that he (Herrick) was in a better position to complete the restoration faster. Frank finally gave in, and Herrick immediately turned over the restoration to Maurice Hovious.

Herrick besides collecting vintage aircraft has recreate special antiqu airplane events. In 2001 he restaged one of Amelia Earhart's intercontinental flights, and two years later organized National Air Tour just like the old Ford Reliability Tours of the 1920s-1930s' Composed of thirty-five vintage arcraft, including the restored Ford Tri-Motor, it traveled 4,000 miles and visited twenty-six cities in the mid-west.

Like most of the planes in the Golden Wings Flying Museum, the Ford Tri-Motor today is not only flyable, but also available for open houses, personal appearances, breakfast events, air shows, and as props in movies.

One of the recent owners of a flyable Tin Goose is Ron Pratte, the American entrpreneur who founded and headed for many years Ron Prette Development Corp., one of the nation's largest concrete foundation asnd wood-forming companies. In 2003 he decided to collect classic autombles and started his collection by buying twenty-two classic cars at one fell swoop. Later on he added antique airplanes to the collection he calls Ron Pratte Collectibles which he keeps in Chandler Stellar Park, Chandler, Arizona.

One of his purchases was a Tin Goose privately owned since 1959 by Dolph Overton, a decorated Korean flying ace who kept it with his Wings and Wheels Aviation Museum. It started flying for Marmer Flying Servuce on the West Coast and Flying Service, it was later sold to K-T Flying Service of Honolulu and was at Pearl Harbor on December 7, 1941 where it was strafed—fortunately with little damage—by Japanese warplanes. Brought back to the mainland in 1946 by a private owner, it was leased by TWA for its 1949 20th anniversary celebration. It then went to an agricultural operator in Idaho and was modified as a sprayer and also as one of the pioneer forest fire fighting air tankers. Johnson Flying Service in Montana flew it for several years to drop Smoke Jumpers and supplies to fire fighters. When Overton purchased it, he put it into the Overton Family Trust, which was created to fund the plane's restoration and facilitate its sale. From time to time Overton took it to different airfields in the South, most recently to the Virginia Aviation Museum, and in 2009 advertised for it sale on the Internet,.

The plane had received from the Overton Family Trust an impressive restoration. A new interior had been installed and the exterior had been completely re-skinned, with most work being performed under the supervision of Master Restorer Bob Woods of Woods Aviation in Goldsboro, NC. Also the wings had been reworked and re-skinned by Maurice Hovious' Hov-Aire company in Vicksburg, Michigan. The landing gear, including the unique Johnson bar braking system, was complete and original, and the original straight-laced wire wheels had tires that were re-sculpted to replicate the correct profile and tread pattern of the period. The wood paneling of the interior has been skillfully recreated. There was no modern avionics or communications gear—just what came with the plane when it was delivered from the Ford factory in January of 1929. Exhaustive efforts were made to ensure originality in every detail with assistance from Tim O'Callaghan of the Henry Ford Museum and American Aircraft Historian Bill Larkins, author of "The Ford Tri-Motor" book. Also assisting were Retired Eastern Airlines Captain Bob Beitel and Retired Admiral Witte Freeman of the Virginia Aviation Museum. Total airframe time is 3110 hours. Total time on the three Wright Whirlwind engines and the propellers is just 56 hours SMOH. The restoration was completed with authentic markings for TAT as a tribute to the historic first air-rail route across the United States by Transcontinental Air Transport in 1929. This was the second time this plane had been used for this purpose, as TWA painted it in a similar fashion for its 20th anniversary national tour in 1949.

Another Ford Tri-Motor flies out of "Planes of Fame" at Valle Airport, an airport about thirty miles from the main airport at Grand Canyon, Arizona. Operated by Grand Canyon Airlines, the successor to Scenic Airlines, which leased the plane from Twin Otter International, Las Vegas, this plane was featured in many TV documentaries and films. For the Mel Brooks' movie, "To Be or Not To Be," it was repainted to look like one of Hitler's private Junkers, swastika and all. Although this Ford Tri-Motor occasionally makes sightseeing trips, more often it flies to air shows. It is also used to train and test airline pilots, who for vanity reasons want to add a Ford Tri-Motor rating to their licenses. This program started in the 1990s by a veteran Uunited Airline pilot Bernie Godlove and was revived by his son Brian Godlove in 2008 after a fifteen year hiatus. The cost of the VFR) Visual Flight Rules certification,which takes three days to complete, is $11,000 plus a fee of $800 to the FAA. Those who take the course receive a manual, group biefing, and four hours of dual flight. Also available for $5,000 is a IFR (Iinstrument Flight Rules} rating.

As for the five non-flyable Tin Gooses these include a 4-AT in the Henry Ford Museum in Dearborn, Illinois. In perfect condition, this is the plane that Admiral Byrd flew over the South Pole. Then another 4-AT in the National Naval Museum of Naval Aviation in Pensacola, Florida. This plane was donated by Dexter Coffin who years ago flew it from time to time, Then, too, there is in the Smithsonian's National Museuem of Air and Space a 4-AT donated by American Airlines. Also there is a 4-AT in the National Museum in Papua, New Guinea. A Royal Australian Air Force plane that crashed in World it lay on the bottom of Myola for more than thirty-five years before being retrieved by a RAAF heliocoptr in what has been described as "a spectacular operation."

And finally a 5-AT in the San Diego Air and Space Museum in San Diego., California. A donation from Skyways Airlines after it was severely damaged by a wind storm in 1992, this plane was worked on for thirteen years by a dedicated group of voluneteers trying to restore it. but the program was stopped when it was realized that once completed, the insurance required to fly the Tin Goose would be too high to maintain,

A special feature of the plane's exhibit it is displayed in the very same rotunda that Ford Motor Co. exhibited it when traveling it around in the 1920s-1930s making sales appearances.

As for the five Tin Gooses being restored, three are being worked on by Maurice Hovious of Hov-Aire, Vicksberg, Michigan. He himself owns two of them, while a third is the Ford Tri-Motor that was in the Weeks Air Museum when it was damagaed by Hurricane Andrew in 1992.

Then there is the Tin Goose that crashed inAlaska in the 1930s and was retrieved fifty five years later. Owned by theAlaska Heritage Museum, it is now in the Golden Wings Museum in Blaine, Minnesota.,

Finallly there is the Ford Tri-Motor being restored by the Tin Goose EAA Chapter 1247 of Port Clinton, Ohio. Maurice Hovious of Hov-Aire is supplying the parts and know-how, and a group of volunteers from the chapter is furnishing the labor.

*   *   *

Chapter Fourteen
The Survivors

Flyable

4-AT 1927 Owner Greg Herrick (Yellowstone Aviation) Golden Wings Flying Museum, Once known as "Niagara" The oldest Ford Tri-Motor in existence

5-AT 1928 Owner Evergreen Vintage Aircraft Inc. Evergreen Aviation Museum, McMinniville, Oregon

4-AT 1927 Owner Greg Herrick (Yellowstone Aviation) Jackson, Wyoming

4-AT 1929 Owner Ron Pratte's Collectible Aircraft Once known as the "City of Richmond" Chandler Stellar Air Park Chandler, Arizona

4-AT 1929 Owner Experimental Aircraft Association EAA AirVenture Museum Oshkosh, Wisconsin, Oregon

5-AT 1929 Owner Kermit Weeks Once Known as the "City of Philadelphia" Fantasy of Flight, Polk City, Florida

5-AT 1928 Owner Kalamazoo Aviation History Museum AIRZOO Kalamazoo, Michigan

4-AT 1928 Owner Sopwith, Ltd. Valle Airport, Valle, Arizona

Non-Flyable

4-AT 1928 Owner Ford Motor Co. Byrd's South Pole Aircraft Henry Ford Museum Dearborn, Michigan

4-AT 1928 Owner National Museum of Naval Aviation Pensacola, Florida

5-AT 1929 Owner San Diego Space and Air Museum San Diego, California

5-AT 1929 Owner The Smithsonian's National Air and Space Museum, Washington, DC

5-AT 1929 Owner National Museum Papua, New Guinea

Under Restoration

4-AT 1928 Owner Kermit Weeks Badly damaged in Hurricane Andrew 1992 Vicksberg, Michigan

4-AT 1929 Owner Maurice Hovious' Hov-Aire Vicksberg, Michigan

4-AT 1929 Owner Maurice Hovious' Hov-Aire Vicksberg, Michigan

4-AT 1929 Owner Alaska Heritage Museun Had name of "Parmigan II" Golden Wings Museum near Minneapolis, Minnestota.

4-AT 1929 Owner Maurice Hovious' Hov-Aire A restoration project in Port Clinton, Ohio, conducted by volunteers from EAA Tin Goose Chapter 1247 in conjunction with Mr Hovious.

*   *   *

Credits and Acknowledgments

This book could not have been published without William T. Larkins. The eminent historian of American aviation history (he founded the American Aviation Historical Society), he has devoted much of his life to promoting—with his writing and photographs—an accurate account of what can only be in his mind a beloved airplane. In 1971, when The Saga of the Tin Goose was first published, Bill Larkins took the trouble to write me a 15-page letter pointing out, in a very kindly way some corrections I should make which I have incorporated in this edition.

My appreciation to William T. Larkins does not end there, since for much of the updating, I depended on his invaluable book on the Ford Tri-Motor which Schiffer Publishing, West Chester, Pa., published in 1992. A veritable Ford Tri-Motor encyclopedia, The Ford Tri-Motor 1926-1992 includes a detailed history of every one of the 200 Ford Tri-Motors manufactured in addition to special sections on the plane's construction, the airlines using them, export and foreign activities, and special events and operations. And beyond that, their book contains hundreds of beautiful photos of current and past Tin Gooses, many of which Larkins took himself.

Special appreciation is also due to Penn R. Stohr, Senior Vice President of Flight Operations for Evergreen International Airlines, McMinnville, Oregon, who gave me much of his time in supplying photos and information on his family's three generations of experience with Tin Gooses.

Also helping out more than they probably realize with photos and/or information were Gerry Walburn, John Burton, and Susan Lurey, EAA, Oshkosh, Wisconsin; Gene Frank, Caldwell, Idaho; Dolph Overton, Smithfield, Virginia; Maurice Hovious, Hov-Aire, Vicksburg, Michigan; Bernice Curton, Kalamazoo Aviation History Museum, Kalamazoo, Michigan; Ray Wagner, Archivist, San Diego Aerospace Museum, San Diego, California; Caroline Roulea, Fantasy of Flight, Polk City, Florida; R. L. McDanel, Virginia Aviation Museum, Richmond, Virginia; John

Ellis, Kal-Aero, Kalamazoo, Michigan; David Stoddard, Scenic Airlines, Los Vegas, Nevada; and Jean Kinner, Twin Otter International, Las Vegas, Nevada.

Last but, as they say, not least, are David Goessling who first gave me the idea of republishing this book and then made me a model of the Tin Goose as a reminder; Robert MacKenzie who gave me the valuable leads that started the project off; Edward Weiss who did the on-line searching of the NEXIS data base to bring me up to date; Isa Carson, long-time associate who did her usual fine job of fact checking; Marion Bomhoff who helped with some of the more difficult editing problems, and Martin Greenberg of Greenberg Consulting who did the final preparation of the manuscript.

Thanks are also due to this book on the Ford Tri-Motor could not have been written without the assistance of numerous individuals, companies, and libraries. Among the companies that graciously supplied research materials and photographs were many of the airlines that originally flew these planes. These included American Airlines, Eastern Airlines, Pan American World Airways, Trans World Airlines and United Airlines. Specific thanks are owed to Althea Lister, Curator, Clipper Hall, Pan American World Airways; Gordon Gilmore and Rose C. Scotti of Trans World Airlines; Laurence M. Grinnell, United Air Lines; Eleanor Camizzi, Eastern Airlines; Anne Vitaliano, American Airlines; and the New York public relations office of the Boeing Company and Lufthansa-German Airlines.

Another company without whose help this book could not have been written was the Ford Motor Company. Particular thanks here should be given to John D. Cameron of the Company's Regional Public Relations Office in New York City, and to Miss Cara L. Benson of the company's Dearborn news office.

Thanks are also due to Ralph Dietrick, President of Sky Tours, Inc., operators of Island Airlines.

The resources of various libraries in the New York metropolitan area were also invaluable in researching this book. These included the Brooklyn Public Library (Grand Army Plaza, Brooklyn Heights Branch, and The Business Library); the New York Mercantile Library; and the Eastern Region Library of the Federal Aviation Administration, John F. Kennedy International Airport. The New York Public Library was also of great assistance. Original research materials not obtainable elsewhere were consulted at the Reference Collection, Forty-second Street and Fifth Avenue, and excellent and historic photographs were obtained from the library's Picture Collection. In addition, the New York Public Library's Wertheim Study was used during the research stages, and for the availability of these facilities special appreciation is due.

Literally dozens of individuals contributed to this story of the Tin Goose. Helping in the beginning stages to get the research off to a good start and then later with the supplying of many rare photographs was Henry E. Edmunds, Director of the Ford Archives, Greenfield Village, Dearborn, Michigan. Others who contributed valuable information were Peter Bruce Walton, Richard B. Miller, Donovan Fitzpatrick, Richard S. Lewis, Edward Starr, Paul Tilles, and Richard Kahn of Aviation Week and Space Technology. Thanks are also due to Barbara Mack, who did such an excellent job of typing the manuscript.

Bibliography

"Bill Stout," *Fortune,* January 1941.

Bombard, Owen. "The Tin Goose," *Dearborn Historical Quarterly,* May 1958.

    An invaluable article written by a Ford public relations official and containing much original material.

"Bushmaster 2000 International Tour Set," *Aviation Week and Space Technology,* September 26, 1966. Caidin, Martin. *Barnstorming.* New York: Duell, Sloan & Pearce, Inc., 1965.

"Colossus of the Caribbean," *Fortune,* April 1931. *Current Biography*

    Brief but excellent biographical sketches of William Bush-nell Stout and Henry Ford are in the 1942 and 1944 editions, respectively.

Damon, Ralph S. "TWA: Nearly Three Decades in the Air." Speech delivered at National Newcomen dinner, 1952.

Davies, R. G. *A History of the World's Airlines.* New York and Toronto: Oxford University Press, 1964.

DeLeeuw, Hendrik. *From Flying Horse to Man in the Moon.* New York: St. Martin's Press, 1963.

"Detroit Da Vinci," *Saturday Evening Post,* December 7, 1940.

Donovan, Frank. *The Early Eagles.* New York: Dodd, Mead & Co., 1962.

Editors of *Air Force Times. Before the Eagle Landed.* Washington and New York: Robert B. Luce, Inc., 1970.

Ellis, Jim. *Billboards to Buicks.* New York: Abelard-Schuman, Ltd., 1968.

Gibbs-Smith, Charles H. *World Aircraft Recognition Manual.* New York: John de Graff, Inc., 1956.

Harris, Sherwood. *The First To Fly.* New York: Simon and Schuster, Inc., 1970.

Hildreth, C. H., and Nalty, Bernard C. *10001 Questions Answered About Aviation History.* New York: Dodd, Mead & Co., 1969.

Ingells, Douglas J. *Tin Goose: The Fabulous Ford Tri-Motor.* Fall-brook, California: Aero Publishers, Inc., 1968.

An excellent book on the Ford Tri-Motor written mostly for aviation history buffs. Amply illustrated, it contains many rare photographs from the Ford Archives. In addition, an entire chapter is devoted to Island Airlines's present use of Ford Tri-Motors at Port Clinton and Put-in-Bay, Ohio.

Kelly, Charles J., Jr. *The Sky's the Limit.* New York: Coward-McCann, Inc., 1963.

Larkins, William T. *The Ford Tri-Motor.* Profile Publications, no. 156. Leatherhead, Surrey, England: Profile Publications Ltd., n.d.

Although only a small pamphlet published for airplane model builders and aviation history enthusiasts, this history of the Ford Tri-Motor is excellent, containing much valuable information and many unusual historic photographs. It also includes an itemized list of the original purchasers of every Ford Tri-Motor manufactured.

Lewis, Peter. *British Aircraft, 1809-1914.* London: Putman & Co., Ltd., 1962.

Mason, H. M., Jr. *Bold Men, Far Horizons.* Philadelphia and New York: J. B. Lippincott Co., 1966.

Nevins, Allan, and Hill, Frank Ernest. *Ford: The Times, the Man, the Company.* New York: Charles Scribner's Sons, 1954.

Nevins, Allan, and Hill, Frank Ernest. *Ford: Expansion and Challenge, (1915-1933).* New York: Charles Scribner's Sons, 1957.

_____. *Ford: Decline and Rebirth (1933-1962).* New York: Charles Scribner's Sons, 1963.

The definitive series of books on the Fords and the Ford Motor Company.

The New York Times. Selected issues from 1924 through 1971.

This newspaper was an invaluable source of considerable information that previous books on the Ford Tri-Motor seem to have omitted or ignored. During the early days of Ford Tri-Motor production, Henry Ford, and to a lesser extent Edsel, were interviewed and quoted as frequently as once a week on not only Ford aviation progress and future plans, but also on aeronautical developments and airline operations in general, both American and worldwide.

"Pan American Airways," *Fortune,* April 1936.

Rolfe, Douglas. *Airplanes of the World: 1490-1962.* Revised Edition. New York: Simon and Schuster, Inc., 1969.

Rolfe, Douglas, and Dawydoff, Alex. *Airplanes of the World.* Revised Edition, New York: Simon and Schuster, Inc., 1969.

Roseberry, C. R. *The Challenging Skies.* Garden City, N.Y.:

Doubleday & Co., Inc., 1966. Stout, William B. *So Away I Went.* Indianapolis: Bobbs-Merrill Co., Inc., 1951.

The autobiography of the eccentric inventor who accomplished the impossible in interesting the Fords in aviation. Somewhat rambling but always interesting, the book also discusses in detail Stout's early days as a "how-to" toy-builder-columnist, and also his later spectacular inventions, such as the Railplane, the Skycar, and the Scarab automobile.

Taylor, Frank J. *High Horizons.* Revised Edition. New York: McGraw-Hill Book Co., Inc., 1961.

An excellent, highly readable, and authoritative book on the history of the development of the leading American airlines.

"The Bushmaster," *Aviation Week and Space Technology,* November 18, 1968.

Toland, John. *Ships in the Sky: The Story of the Great Dirigibles,* New York: Henry Holt and Company, Incorporated, 1957.

"Was There Ever a Sales Story like This?" *Magazine of Business,* November 1927.

Wigton, Don C. *From Jenny To Jet: Pictorial Histories of the World's Great Airlines.* New York: Crown Publishers, Inc., Bonanza Books, 1963.

About the Author

A graduate of Johns Hopkins University, David A. Weiss began a successful literary career after World War II, writing articles for magazines such as the Reader's Digest, publicity features for Universal Pictures, and several books, including the original version of the Ford Tri-Motor book and The Great London Fire, the colorful story of the 1666 conflagration that devastated London, and in the late 1960s he founded Packaged Facts, Inc., one of the nation's leading market research companies. Mr. Weiss is a member of the Knickerbocker Field Club.the Baker Street Irregulars, and the American Revolution Round Table.(New York Chapter).